CANCER

AS INITIATION

CANCER
AS INITIATION

Surviving the Fire

A GUIDE FOR LIVING WITH CANCER
FOR PATIENT, PROVIDER,
SPOUSE, FAMILY, OR FRIEND

Foreword by
Richard K. Orr, M.D.

Barbara Stone

OPEN COURT
Chicago and La Salle, Illinois

First printing 1994
Second printing 1995

Printed and bound in the United States of America.

Library of Congress Cataloging-in-Publication Data

Stone, Barbara (Barbara E.), 1948–
Cancer as initiation : surviving the fire, a guide for living with cancer for patient, provider, spouse, family, or friend / Barbara Stone.
p. cm.
Includes bibliographical references and index.
ISBN 0-8126-9273-X (cloth).—ISBN 0-8126-9274-8 (paper)
1. Stone, Barbara (Barbara E.), 1948—Health. 2. Breast—Cancer—Patients—United States—Biography. 3. Breast—Cancer—Alternative treatment. I. Title.
RC280.B8S74 1993
362.1'9699449'0092—dc20
[B] 94-36451
CIP

This book is dedicated to my daughter
and to all of the other women in the
world who are at risk for breast cancer.
It is also dedicated to the
hundreds of thousands of women who
are presently facing breast cancer
and to the millions whose lives have been
claimed by this disease.

CONTENTS

Forward xi
Preface xv

1 Rite of Passage 1

Framing the experience of breast cancer as a rite of passage from the former state of nurturing others to beginning to learn how to nurture and heal my own body, an initiation into mature womanhood.

2 Newlyweds 13

Getting remarried just three months before finding the lump. Emotional ambivalence about the loss of independence. Adjusting to the new relationship.

3 Finding the Lump 19

Images of danger and death coming in dreams and synchronicities (including the suicide of a colleague's cousin) urging me to do the self-exam, which found the lump.

4 Snake Dreams 23

Viewing the snake as an image of cancer. Examining the death/transformation ambivalence of snake symbolism in dreams from a year before diagnosis to the week before finding the lump.

5 Diagnosis 37

Mounting tension during the month of waiting to see whether the lump was malignant. Gradually shifting my attitude from denial of having cancer to bracing myself for the truth, which I already had in unconscious knowledge. Getting the diagnosis of cancer the night before surgery to remove the lump.

6 Change of Heart 50

Disbelief and emotional turmoil during the sixteen hours between diagnosis and surgery. Contacting parents, children, former husband, and friends to ask them to pray for me. In prayer, feeling my heart touched and my soul being healed. Intuitively knowing that with my soul healed, my body would also heal.

7 Surgery 54

Fear of surgery overshadowed by deep inner joy from the experience of soul healing the previous night. Floating through the hospital stay on an emotional high from this healing.

8 Staging the Cancer 58

Crashing when home from the hospital. Being hit with the emotional impact of cancer. Nightmares about having breast cancer. Knowing I had to write my doctoral thesis on surviving breast cancer, including my personal experience. Feeling relief that none of my lymph nodes had any cancer, putting my case into Stage I, the most treatable type of cancer.

9 To Chemo or Not to Chemo? 74

Health recommendations from Deepak Chopra's Maharishi Ayurveda Health Center. Struggling with the decision of whether or not to get chemotherapy. Not wanting to lose my hair. Then reframing the experience as a shamanic initiation, with chemotherapy and hair sacrifice necessary to increase my chances of survival.

10 Beginning Radiation Therapy 86

Feeling an unexpected physical toll from radiation, including periods of depression. Changing relationship with death: sometimes fearing it, sometimes longing for it, yet wanting to finish my work here first.

11 New Vision 97

Conflicts with my husband resolved by my mentor reframing our marriage as partnership at a soul level. Framing the passage through cancer as part of a larger initiation, which included learning how the mind-body connection could be used to promote healing.

12 Ups and Downs 110

Difficulty getting through the side effects of the radiation treatments. Difficulty coping with routine stressors because of lowered energy level and mood swings. Confronting my internal image of the source of the cancer, "The Black Inky Thing."

13 Starting Chemotherapy 131

Physical and emotional effects of the first chemotherapy treatments. Intuitively sensing that the cancer was gone from my body after the second round of chemotherapy.

14 Kirlian Photography 149

Kirlian photographs of my energy pattern during treatment, registering changes in my energy field from radiation, chemotherapy, and the supplementary treatments I added to traditional Western medicine. Also showing changes in my energy field from stress, expressing emotion, and repressing emotion.

15 Healers 173

Healing from the emotional scars of being molested, with significant improvement in physical energy afterward. Getting new insights on using the mind-body connection to heal my own body during a group healing meditation session.

16 Finishing Chemotherapy 184

Going into crash menopause after the fourth chemotherapy session. Having more difficulty with nausea during the final two chemotherapy sessions, but feeling great joy at completing the treatments.

17 Losing My Hair 206

Describing techniques and alternative measures used to prevent loss of hair from my head. Discussion of the placebo effect and the tool of visualization.

18 Baby Dreams: New Life 215

Tracking dreams about pregnancy, childbirth, and babies over a four-year period, from before meeting my husband, through cancer treatment, to reorganization of my life after chemotherapy was over. Birthing a new aspect of "the Self," uniting conscious and unconscious parts of the mind to deepen the personality; shifting from the mother stage of womanhood into the mature "Wise Woman."

Appendix A: Cranial Osteopathic Philosophy 239

Appendix B: Kirlian Photography Theory 241

Appendix C: Kirlian Apparatus 246

Appendix D: Acupuncture 250

Appendix E: Funding for Breast Cancer Research 253

References 257

Index 263

Audio Tape "Initiation" © and Workshop Information 267

FOREWORD

The popular press had become obsessed with breast cancer. Perusal of magazines targeted for women show repeated efforts to "educate" their readers, often resorting to gross oversimplification or even subtle mistruths. Despite the number of magazine articles and television "sound bites," there are very few in-depth discussions of the reality of breast cancer from a patient's viewpoint. Barbara Stone has written an excellent account of her own personal experience as a breast cancer patient. This book shares some features with others of its kind. Barbara's experience with the fear associated with the diagnosis of breast cancer and the discomforts of surgery and adjuvant treatment could be "every woman's story." Obviously there are variations between people, but few would dispute her frank discussions about the suffering a breast cancer patient endures with our current treatments. Beyond these chapters, however, it is clearly Barbara Stone's personal story, reflection, and vision.

Most patients, at one time or another, ask their physicians, "Why did I get this cancer?" Occasionally, we have some sort of an answer if the patient has engaged in very high risk behavior (for example, a lung cancer patient who smoked) or has a very strong history of cancer in the family. Usually, as in Barbara's case, we have no idea. I am certain that most patients gradually forget the question, or attribute it to "God's will." Ms. Stone takes the question to a logical extreme, and describes her cancer as an "initiation rite." She has drawn on beliefs of other cultures as well as dream analysis and other techniques to create a very reasonable and readable account of this initiation. She has obviously survived this rite, and as such has become stronger for it. Barbara Stone articulates a concept that I have seen in many other patients who put their cancer behind them and emerge with new life purpose, direction, and vigor. Although few of my other patients use Barbara's lexicon, I have no doubt that the phenomenon is real and worthy of scientific study. I thank Barbara for bringing these ideas and concepts to print.

I am convinced, as are many oncologists, that there is clearly a mind-body connection. Patients who somehow achieve an inner peace and sense of purpose seem to do better than those who give up. This issue is finally getting the scientific study it deserves. Nevertheless, it is important to realize, and Ms. Stone clearly points this out, that at the present time it makes little sense to abandon traditional medicine completely. I would advise any patient, cancer or otherwise, to seek out an empathetic physician who will discuss all of the available treatments and their risks/benefits. Barbara Stone is beating her cancer because of her psyche, but also, we both feel, because of the medical treatments she received. I believe that there are appropriate times to refuse medical treatments, but any such decisions should be based on a firm foundation of knowledge.

As a surgical oncologist, I cannot resist a few editorial comments. Dr. Susan Love refers to the crudeness of modern day cancer therapy as "slash, burn, poison." I could not agree more. As physicians, we must strive to eliminate suffering, and to constantly decrease the morbidity of our treatment. Few would argue that the lumpectomy and axillary node dissection that I performed on Barbara Stone were as bad as the radical mastectomy that my surgical teachers would have performed thirty years ago. Nevertheless, as Barbara clearly shows, my operation was hardly trivial. We need to continue research to sort out which patients need which operations/chemotherapy/radiation and to stop indiscriminate treatments unless the patient is fully aware of the risks and benefits in her *particular* case. Research must continue on the appropriate assessment of the harm we as physicians do, and the effect on quality of life for the patient. Parenthetically, had Ms. Stone been diagnosed in 1994, I would have been very clear about the optional nature of the axillary dissection in her case. In view of her young age, we would have probably performed the operation after our discussion, but an older post-menopausal woman might have made a different choice.

I have little doubt that major improvements in breast cancer care will come not from practicing physicians but from basic scientists elucidating the molecular causes of cancer. Intervention must be at a much earlier level than is foreseeable with our current technology. Until such a time occurs, however, we need to continue to support research into the mind-body connection in cancer patients. I am currently involved in a research study which will look into a variety of "non-medical" interventions and their effect on patient well-being, immune function, and ultimately, survival. Such research is vital, and must continue. Until we can prevent cancer, or at least stop it in much earlier phases, we must continue to identify ways of making cancer treatments more bearable and less toxic.

I congratulate Barbara Stone as she continues to win in the struggle with her former cancer, and I congratulate her on writing a fine and important book.

Richard K. Orr, M.D.
Director of Surgical Oncology
Fallon Health Care System

PREFACE

This book represents my reality, my truth. Since I believe each person creates his or her own reality, your reality may be different. My theories have grown out of my direct personal experience and may or may not be shared by my colleagues. I am speaking only for myself.

If you find that a piece of my truth does not fit into your truth, I invite you to let go of that thread of the fabric I have woven here. Maybe you will be ready for it later, or maybe seeing why it does not fit into your truth will help you define your own reality more clearly.

I am deeply grateful to the advances of Western medicine, which have saved my life three times so far. The first time was when I had pneumonia as a baby; the second time was when my appendix ruptured at age 16; and the third time was by removing the cancer that was in my breast. I cherish the scientific method, the basis of Western medicine, which has brought our medical knowledge to the sophisticated levels we enjoy today. However, even with the help of science, technology, and the scientific method, the death rate from cancer continues to climb.

Being diagnosed with cancer made me reach out further to try to understand this world I live in and to understand the miracle of the way our physical bodies work. It opened up my mind to consider new ideas, which might be beneficial to my health, even though some of these ideas are difficult to test with the scientific method because so many variables are at work. In particular, I drew my circle of cancer treatment bigger than just Western medicine and included ideas from Asian medicine, which has accumulated the wisdom of experience over thousands of years and which views the body as an energy system.

Quantum physics has shown us that matter, which appears to be solid, is really mostly empty space with tiny particles of energy in orbit. We still do not understand all of the theories of the physical universe. Nor do we fully understand the mysteries of healing. Perhaps the missing link in our cancer survival search is not a drug, but something else in the emotional, mental, or spiritual realm that we have not thought of yet.

I tried out many avenues to improve my health in addition to the standard medical treatments. Since my body is extremely sensitive, my system reacted strongly to chemotherapy. However, this same sensitivity also helped me to know very quickly whether or not a new herb or alternative treatment method I tried was helping me to feel better. I carefully watched the responses of my body to each treatment.

My search for survival tools also included Kirlian photography, a method of measuring energy, which was developed to a high state of art in the Soviet Union, behind the iron curtain. In the 1970s, Soviet researchers were unwilling to disclose all the secrets of this technology to Americans. But perhaps now that the relationship between our two countries is healing, we can work together with the common goal of making progress in understanding and treating cancer. Each Kirlian device is different, and our Kirlian technology has been crude compared to theirs. I hope that American research will allocate time and funding to do scientific research on the capacity claimed by some former Soviet bloc countries that Kirlian photography can make an early, accurate diagnosis of cancer.

I am not a medical doctor and do not presume to advise cancer patients on what course of medical and adjuvant treatments to pursue. Each decision must be made by the individual patient together with the physician, based on the situation and the information available. In this book I report how surgery, radiation, and chemotherapy helped me and how the extra interventions I added eased the side effects of these medical treatments. *Clearly, I am NOT advising cancer patients to abandon medical treatments and only do alternative therapies.*

I have my own particular labels I put on the forces of love and healing in the world. Some of my favorite names for love energy are "Mother-Father God," "Lord Jesus Christ," and "Kwan Yin," the Chinese goddess of compassion. But whatever name connects your spirit to your innate wholeness is fine. The force of Life and Love is the same no matter what label holds more meaning for you personally.

ACKNOWLEDGMENTS

I am deeply grateful to all of the wonderful people who supported me through the process of diagnosis, treatment, and recovery from breast cancer. I send heartfelt thanks to all of my family, friends, teachers, classmates at Pacifica Graduate Institute, colleagues, and health care providers who surrounded me with their love and prayers to pull me through the passage. The list of people who get credit for helping me is so long that space does not permit me to name each one individually; my heart sends deepest gratitude to every one of you.

A special thanks goes to my husband, who stood by me each step of the way even though we had been married only four months when I received the diagnosis of breast cancer. His unfailing love and support placed a solid rock underneath me, which held me while I faced the fire. Then he tended the home fire with patience and care during the time when my attention was fully absorbed in the process of writing about the experience.

I also wish to thank the staff of Open Court Publishing Company for their warm cooperation in making this book possible. In particular, I thank Louise Carus Mahdi for her encouragement all through the writing process and Kerri Mommer for her perceptive editorial help.

I am also grateful to the persons who read early copies of the book and gave their input and encouragement. In particular, I thank Edith Sullwold, Ph.D.; Richard Orr, M.D.; Kathleen Tamilio-Awed, Lic. Ac.; Darren McDonough, M.A., MFCC; Bonnie Siegel [Rebecca Boni], M.A.; Karen Blasdell, Ph.D.; and my mother, Rosa Stone, Ed. D.

A special thanks goes to my friend Ruth Anthony, an author, professor, and Jungian analyst who read the manuscript in early 1994 while she was going through chemotherapy for liver cancer. Ruth loved the book and found it resonated deeply with her own truth. She felt that the message from Karpinski in Chapter 11 was directed right to her. She said this book was the one she wanted to write next, to totally "come out" about the spiritual being that she is. Ruth died from liver failure on March 28, 1994, two weeks before her latest book, *Rapture Encaged—The Suppression of the Feminine in Western Culture*, was released, published by Routledge of London under the name Ruth Anthony El Saffar.

NOTE ABOUT THE SYMBOL OF THE DREAM CATCHER

Native Americans have always known the importance of watching dreams, listening to them, and paying attention to the messages dreams bring to guide their lives. They hang a "dream catcher" over the cribs of their infants and say that the screen part of the symbol catches the "good dreams," the hole in the middle lets the "bad dreams" through, and the feathers attached at the bottom bring the good dreams down to the baby. Through the use of the dreamcatcher, Native Americans eliminate their fear of listening to their deeper selves and are able to open up to the wisdom that comes from the unconscious part of the mind.

Jungian psychology teaches the same wisdom but in different words. Jung said there are no "bad dreams"; on the contrary, a nightmare may have the most important message for us, to warn us of danger approaching. All dreams are to be listened to and honored. The main stumbling block to opening up to the content of our dreams is the fear of what messages might come through. The beauty of the dream catcher symbol is that it gives visible protection from this fear, enabling the dreamer to truly open up to catch the flow of life energy that comes through dreams.

> *"Dreams come to tell us something about our lives that we are missing."*
>
> —James Redfield

> *"Dream symbols are the essential message carriers from the instinctive to the rational parts of the human mind, and their interpretation enriches the poverty of consciousness so that it learns to understand again the forgotten language of the instincts."*
>
> —Carl G. Jung, 1964, p. 37

CHAPTER 1

Rite of Passage

Perhaps cancer is an experiment, an experiment in the creation of greater personality. (Lockhart, 1983, p.66)

Introduction

Fear gripped my soul when I discovered a marble-sized lump hidden in my left breast during a self-examination in July of 1991. My first thought was, "Cancer equals death," and I thought I would die if the lump turned out to be malignant. This book is the story of the journey from that initial point of panic, thinking that cancer was a death sentence, to the gratitude I now feel for the opportunity cancer gave me to leave behind my old manner of life and step into a richer, fuller way of living.

As I journeyed through the diagnosis, surgery, radiation, and chemotherapy, the terror of death that started this transformation turned into love for my life. As my body was being healed from the cancer, my soul began the ongoing process of peeling away the layers of emotional wounds barricading my heart, learning how to open up to love others *and myself* more fully. Instead of bringing death, cancer brought me new life.

I asked many questions along the way. I wondered why this cancer had invaded my breast and not some other part of my body. Why did I get breast cancer so young, at the age of forty-two? Why did the diagnosis come just four months to the day after I had re-married? Why did it come just as my last child was leaving home? Why has so little research been done on this specific form of cancer? And why has this disease become an epidemic?[1]

Stages of Womanhood

Cancer initiated me into a different stage of womanhood. The three main phases in a woman's life have been called by many different names (Walker, 1983) and correspond to the different phases of the moon:

WAXING MOON	FULL MOON	WANING MOON
Youth	Middle Age	Old Age
Premenstrual	Fertile	Post-menopausal
Child	Bride-Wife	Widow
Maiden	Adult Woman	Wise Woman
Little Girl	Mistress of a House	Grandmother
Virgin	Mother	Empty Nest/Crone
Naivety	Life Experience	Deep Wisdom

Cancer jerked me out of finding my primary identity in the second stage of womanhood and abruptly plunged me into the third stage, since the chemotherapy induced crash menopause. Simultaneously, my last child left home, leaving my nest empty and terminating the "active duty" part of my mothering role. But an even deeper change in my life was that the trauma of going through a life-threatening illness challenged me to grow.

Surviving the Fire

At the time cancer arrived, I had been balancing my primary job of being a clinical social worker four days a week and doing psychotherapy in a community health center with making pottery on my days off. As a potter, I took raw clay, formed it into a shape, and then put the pot into a very hot fire, from 1700° to 2200°F. Inside the kiln, the pot was

---

[1]Breast cancer will kill 46,000 American women this year. The rate of this disease, which is the number one killer of women in my age bracket (Lauder, 1991), has risen from its risk in our population fifty years ago of affecting one woman out of every twenty to the present risk rate of one in eight (Associated Press, 1992). My home state of Massachusetts is the first state to officially declare breast cancer an epidemic.

transformed from mud into beautiful, useful stoneware by burning out the impurities in the clay body and heating the clay almost to its melting point so that the clay turned into stone. The hotter the fire, the stronger the pot came out; however, sometimes the pots cracked or exploded in the kiln.

Cancer was my initiation into the "Wise Woman" stage of life, but the fire inside of the surgery-radiation-chemotherapy kiln was very hot, and at times I feared that I would not survive the transformation process and might explode, losing my physical body. I also feared that I might come out of the fire cracked, with my physical health permanently damaged or with my physical body functioning flawed in some way.

Facets of Initiation

Anyone who has faced a serious form of cancer and has survived has probably been touched by this experience, but initiations can also come in many other forms of life trials and challenges.

> Every dark thing one falls into can be called an initiation. To be initiated into a thing means to go into it. The first step is generally falling into the dark place and usually appears in a dubious or negative form—falling into something, or being possessed by something. The shamans say that being a medicine man begins by falling into the power of the demons; the one who pulls out of the dark place becomes the medicine man, and the one who stays in it is the sick person. (von Franz, 1972, p. 64)

This journey through cancer had many of the hallmarks of traditional initiation rituals (van Gennep, 1960), including a test of emotional and physical strength, proof of the ability to withstand physical pain, scarification, and a time of separation from the mainstream of life during which a person goes inside her own heart to find the meaning of her life. The old life then ends in a symbolic death, and the initiate is "reborn" into a new psychological and physical state. She enters back into the community with a new role in society, with a deeper connection to the wisdom in her own heart, ready to share what she has learned for the benefit of the entire community.[2]

---

[2]Cultures all over the world have ritualized transitions through different stages of life with initiation ceremonies, and the need for a ceremony or a "rite of passage" to mark a transition is universal. Mythologist Joseph Campbell described how tribes helped the individual flow

Native Americans mark the passage of a young person from youth into adulthood with a practice that is today called a "Vision Quest." In this initiation, an individual goes away from the tribe out into the wilderness carrying nothing except a buffalo robe, and then the initiate fasts and meditates until clear images come of the purpose of his or her life. The person might also get a new name or a song before coming back to the tribe and telling the story of what happened, sharing the vision received with the tribe for the good of the entire community (Foster & Little, 1987). This Vision Questing can be repeated at any turning point in life when inner guidance is needed, finding answers to major life questions on the inside instead of trying to find the answers on the outside, either through prescribed religious forms or by fulfilling expectations others have set for one's life.

Cancer has been a psychological Vision Quest for me in which facing cancer threw me into an emotional wilderness filled with many physical dangers. To fight the cancer, I let my body be cut and stabbed by the knife in surgery, be exposed to high levels of X-rays in radiation therapy, and be poisoned by chemotherapy. Through the quiet times of meditation during my journey, I discovered that the reason I came to earth was to learn secrets of healing, and this book shares that vision with you, my readers, my neighbors, my community. The process of telling my story has helped me heal, and I hope that others might also benefit from listening to this story of one woman's journey through breast cancer.

Is Surgery an Initiation?

Like Native Americans, most other cultures have prescribed rituals to help a person get from one stage of life into the next one more smoothly. However, my American white Anglo-Saxon Protestant (WASP) culture has no traditional ceremony to honor a woman's crossing the threshold from motherhood into menopause. Swiss psychiatrist Carl Jung taught that if we do not plan a passage consciously, then we have to pay for that fact by doing it unconsciously.

---

through these life changes, which could be frightening, by marking the transition with a tribal ceremony, a form that preserved the continuity of the society, giving the tribe a sense of immortality even as individuals passed through the life stages of birth, adolescent initiation, marriage, middle age, old age, and death. "Generations of individuals pass, like anonymous cells from a living body; but the sustaining, timeless form remains" (Campbell, 1968, p. 383).

Since my culture did not plan an initiation for me into the third stage of a woman's life, I chose to create a transition out of the illness I went through.

Possibly surgery, with its initiation element of scarification, may be an unconscious way for people in my culture to cross into the final stage of life, which reflects deepened wisdom. Few reach their grave having escaped the surgeon's knife!

Going through major surgery includes some features that are essential elements of traditional initiations. For example, family members rally around the patient going through an operation. The patient is separated from the community by being placed in the seclusion of the hospital and is ministered to by the priest-doctor in a procedure intended to cleanse the person from the "evil spirits" of physical illness. The patient then re-enters society with a visible scar marking the whole ordeal.

Scarification

When I first found the lump, I feared my breast would be mutilated or cut off, as doctors will almost invariably prescribe surgery for each of the 180,000 new cases of breast cancer, which will occur in American women this year (Love, 1993). The radical mastectomy, which leaves huge scars on the chest extending into the armpit, is often the procedure recommended to treat this form of cancer (Moss, R., 1991).

Surgery leaves scars, and *scarification,* encompassing the ability to withstand physical pain, is a test of strength in many initiation rituals. Scarification is within "the category of all practices of the same order which by cutting off, splitting, or mutilating any part of the body modify the personality of the individual in a manner visible to all" (van Gennep, 1960, p. 71).

The popular Steven Spielberg film *The Color Purple* (1985) showed a suspenseful African adolescent initiation ceremony in which Adam's passage from boyhood into manhood was marked by gashing his cheek in a ritual scarification pattern as a permanent, visible sign that he then belonged to the tribe. An African classmate in graduate school told me about his adolescent initiation circumcision, done with crude tools and without anesthesia of any kind. After the ceremony, he lived secluded in the bush, together with two other initiates, until their wounds had healed. He commented, "After that experience, I *knew* without any doubt that I had crossed over from being a boy into becoming a man!"

A beautiful book called *The Circle of Life* has collected photographs of human rituals from all over the world and contains a dramatic picture of a Kau woman in the Sudan having her nude torso marked with a pattern of welts from the breastline to the pubic bone as part of a life transition, marking her new status (Cohen, 1991, p. 92). When a young man of the !Kung hunter-gatherer tribe of the Kalahari desert made his first kill, his chest, back, and arms were scarified to give him power as a hunter, and in the Tlingit tribe of southeastern Alaska, the traditional puberty rites of a young woman passing into adulthood included tattooing and cutting of flesh (Fried & Fried, 1977).

In my cancer treatment, my chest was tattooed with the points for radiation therapy, and my chest and armpit were scarified in the surgery. However, instead of becoming a better hunter, I gave up eating all animal flesh!

Need for Ritual

Because the "empty nest syndrome" of the third stage of womanhood may be experienced as a tremendous loss, an inner death, the need for a ceremony to mark the transition into this stage of mature feminine wisdom may actually be more intense than the need in adolescence (Henderson, 1964), where getting a driver's license, being able to vote, or getting married may mark the passage from youth into adulthood. The transition from the second stage to the third stage is a rocky course, full of dangers in the "betwixt and between" state of liminality (Turner, 1987), the threshold in between giving up the old form of life as mother and taking on the new role of wise woman.

Whenever a person moves from one state into another, the old state dies to make room for the new one, like the grain of wheat that must die to sprout a new plant. When getting a driver's license, the former childlike state of dependency dies, in which one had to rely on others to get around in the world. And when a man takes on the responsibilities and joys of marriage, his bachelorhood dies to make room for his new role as husband.

> We die and are reborn so many times. The passage from one state of being to another occurs not just at birth and death, but over and over throughout life. (Cohen, 1991, p. 229)

Wholeness: "The Self"

Many initiation rites are marked by a symbolic death in which the conscious personal control the initiate has over his or her life is relinquished to control by a higher level of maturity within that same person, a wiser, inner force which seeks full realization of one's potential. Carl Jung named this energy, which seeks to bring the person wholeness, "the Self" to free it from the limitations of any specific religious framework (Jung, 1974). Since I grew up in a Christian home, images of "the Self" sometimes came to me in the form of Lord Jesus Christ and Mother Mary; however, for a person raised in a different background, "the Self" might have been imaged in different ways, such as "The Great Spirit," a mandala,[3] Kwan Yin, Allah, the Divine Mother, or simply a figure radiating love and light.

For example, the great Eastern symbol of the psychic unity of "the Self" was Buddha, who was originally represented as a twelve-spoked wheel. Buddha was not represented in India as a human figure until its culture had some contact with Greece (von Franz, 1975). The pure, unspoiled culture of the Naskapi Indians from the forests of the Labrador peninsula called this inner center "the Great Man" and cultivated a deeper relationship with this inner companion by paying attention to their dreams, which gave them the guidance they needed for survival by helping them foretell the weather and guiding their hunting (von Franz, 1964).

Images of Death

Letting go of conscious control of one's life may feel like death to the ego, which might complain vigorously about the transformation's symbolic death phase that precedes regeneration at a higher level. In fact, many initiations in primitive societies are analogous to the rites used for the transition of death (Zoja, 1989).

A diagnosis of cancer immediately raises images of death, since the overall five-year survival rate for all forms of cancer combined is only 50%, and the apparent increase in the percent of women who survive five years after diagnosis of breast cancer from 63% in 1960–1963 to 75% in 1979–1984 is largely due to a manipulation of figures.

---

[3]A mandala is a Buddhist or Hindu symbol of the universe and may include a circle within a square with a deity on each side. Any circular or square pattern that radiates out from a central point may be considered a simple mandala.

The arbitrary timeline is to pronounce a woman "cured" if she is not dead five years after diagnosis of breast cancer. Since mammograms are detecting cancer in earlier stages, more time is often needed to progress to the end stages of the disease (Moss, R., 1991).

The very word "cancer" puts the fear of death into the heart, and death itself is the final initiation into the world beyond. The challenge the initiation of cancer presented to me was to let my old attitudes die and to let a new life begin within my heart. The fear of dying from cancer—the fear of leaving the planet at an early age through a slow, disfiguring disease—gave me a strong motivation to change my life, because I could not be sure how much time I had left. The paradox of healing is that confronting death and turning this enemy into a friend is a prerequisite to truly enjoying the precious gift of life in this very moment, which is the only time we ever have.

Shift in Perspective

As I learned to stop running from my fears and took time out from the rat race pace I had imposed on my life, I often wondered just exactly how the cancer had gotten into my body. In the beginning, I felt a passionate need to know what had caused this strange disease. I pondered the words of Russell Lockhart (1983) that cancer was related to denying something in oneself, "something of one's psychic and bodily earth not allowed to live, not allowed to grow. Cancer lives something of life unlived" (p. 54). I wondered where I might have been blocked in my conscious growth so that the growing had come in the form of cancer, an unconscious growth process that threatened my life.

I finally decided to grow as much as I could in the areas of developing my talents, understanding myself, enjoying my time on this earth, and getting as comfortable as I could inside of this physical body, which is housing the rest of me at the moment. I eventually gave up trying to figure out what had caused the cancer. Maybe the cause was simply increased levels of environmental pollution. Instead, I decided to focus all my energy on *surviving* the illness.

Breast Cancer Research

I had none of the known risk factors for breast cancer except that menstruation began fairly early, at age eleven. But 80% of the other women who get this disease are in the same boat, with no risk factors

(Wiley, 1992), a fact that hints at the possibility that we do not really know what causes this particular form of cancer.[4]

In the past, breast cancer has received relatively little attention,[5] perhaps because only 1% of breast cancer cases are diagnosed in men (van Tets, Leenen, Roukema, & Pipers, 1990). Most of the people who made funding decisions were not themselves at high risk for the disease.[6]

However, with the rise of the women's movement, breast cancer has been getting more publicity, and hopefully this change will result in more actual research. Most of the money allocated for breast cancer is presently spent on studying chemotherapeutic drugs for treatment and getting mammograms for early detection rather than focusing on finding ways to prevent malignancies from beginning to form in the first place.

Along the way on my healing journey, I have formulated my own theories about what physical and emotional factors might encourage breast cancer so that I could reduce these influences in my life-style. I am making conscious choices to improve my health and strengthen my immune system, thus lowering my risk factors for growing another tumor.

The emotional blockages I found when I took a careful look inside may not have been the cause of the cancer, but they were impeding the healing process and needed to be removed so healing energy could circulate freely within my physical body. I needed all the strength I could find just to survive the rigorous treatment for this disease.

---

[4]Only about 10% of the 180,000 cases of breast cancer diagnosed each year stem from hereditary defects. In 1993 researchers were looking for the gene that causes breast cancer. They believed that flaws in its genetic pattern were responsible for at least half of the inherited cases (Cowley, 1993, p. 46). Genetic material is encoded on strands of DNA wound into 23 pairs of chromosomes in each cell. Premenopausal breast cancer was suspected to be linked to a region on chromosome 17, which is 50 million base pairs in length. Scientists narrowed this area down to 300,000 base pairs, 12 genes. Scientists made the comparison of the chromosomes being like 23 sets of encyclopedias. The search was down to "part of one volume, looking for a misspelling" (Cowley, p. 49). In 1994 the gene that causes breast cancer was finally discovered and was dubbed BRCA1 (for Breast Cancer 1).

[5]A review of medical literature for 1985–1992 yielded 200,983 articles written on drug therapy/chemotherapy, but only 219 of these articles were on the treatment of breast cancer.

[6]For a discussion of the funding of breast cancer research versus prostate cancer research, see Appendix E.

Nurturing Others

Breast cancer is linked to a woman's sexual functioning[7] and attacks the part of her body that is the sexual symbol of nurturing others, since breasts are used by a mother to give nourishment to her baby. Breasts could thus be seen as a symbol of taking care of others, producing milk for others to drink, since a woman does not drink her own milk.

As part of healing from cancer, I had to examine my attitudes toward nurturing. I found that while I was usually ready to nurture others, I did not often put my name very high up on the list of people I needed to be taking care of. To get through the initiation of breast cancer, which coincided with my nest emptying and menopause, I needed to shift more attention toward attending to my own needs.

The sacrificial focus of a woman's initiation into the third stage of womanhood is different from a man's point of sacrifice as he faces mid-life crisis or initiation into becoming a "Wise Old Man." A woman's initiation has an initial trial of strength that leads up to an inner sacrifice the woman must make to bring the transpersonal energy of "the Self" into her life.

> This sacrifice enables a woman to free herself from the entanglement of personal relations and fits her for a more conscious role as an individual in her own right. In contrast, a man's sacrifice is a surrender of his sacred independence: he becomes more consciously related to woman. (Henderson, 1964, p. 132 & 134)

In other words, to be more complete, a man needs to give up some of the fierce independence our culture fosters in males and to become more conscious of his relationships with other people. In contrast, a woman, who has been conditioned to be highly aware of her relationships with others, needs to become more conscious of her

---

[7]Breast cancer is different from other forms of cancer because it is linked to the hormone estrogen. Hormones regulate all aspects of reproduction, and the estrogen produced by the ovaries leads to the release of a mature egg into the fallopian tube, the beginning of making a baby who could feed at the breast. Estrogen is a vital hormone, which also has other functions, including preventing cardiovascular disease and osteoporosis, but excessive levels of estrogen over a lifetime seem to correlate with increased risk of developing breast cancer. Higher levels of estrogen come with early onset of menstruation and late onset of menopause. "It seems that the more periods a woman has over her lifetime, the more prone she is to breast cancer" (Love, 1990, p. 144). "The earlier a woman's menarche, the higher the levels of estrogen produced in her body. Estrogen production . . . seems to be set at a higher rate by the body's hormonal regulatory system in women with early onset of menstruation" (Gorbach, Zimmerman, & Woods, 1984, p. 37).

independence, more freed up from the responsibilities towards others so that she can figure out who she really is and develop a better relationship with herself. Caring for self and caring for others need to be in balance in both sexes.

Nurturing My Own Life

Perhaps my attitude toward taking care of myself was diseased, and the location of the tumor in the breast, the organ of nurture, was no accident. The malignancy in my breast screamed out to me to change my pattern of nurturing everybody *except myself.* I needed all of my energy focused on my own healing to get through the ordeal of cancer. Instead of attending to the needs of those around me, I had to focus on learning to love and care for my physical body and my emotional well-being.

The necessary medical treatments I chose of chemotherapy, radiation, and surgery took a heavy toll on my physical body and made me feel depressed at times, so I added some alternative healing measures to help my body heal faster and to strengthen my immune system. I tried out everything I could find to improve my physical and emotional health: meditation, yoga, prayer, visualization, osteopathic cranial manipulation, acupuncture, sauna baths, vitamins and herbs, studying my dreams, and diet and life-style changes.

Writing about the process also helped me to heal. Keeping a journal is an easy and inexpensive way for anyone to promote emotional health since it gives a person a safe place to express feelings. The night I came home from surgery, my dream world told me that I was to write my doctoral thesis on the topic of breast cancer and to focus all my energies on my own healing, documenting everything I went through, using my direct experience of breast cancer as the entry point into writing about this disease. In meditation, I received an insight to title the book *Cancer as Initiation.* I hope that through reading this book, other women may grow to become more loving to themselves and might go through the initiation into the "Wise Woman" phase of womanhood *without* getting cancer or chemotherapy.

The Healing of My Soul

Cancer has been my teacher. In searching for both a personal and a collective meaning to this experience, I discovered that the part of me that needed healing the most was not my body but my soul. Through telling my story, including sharing my journal entries and dreams, this book traces the steps from my emotional state before the cancer through the diagnosis and treatment to the beginning of the ongoing process of the healing of my soul, the real story I want to share. Cancer initiated me into a new way of *being* in the world.

CHAPTER 2

Newlyweds

The story begins two years before the diagnosis of breast cancer, when I first met my second husband Soli in November of 1989. At that time I had been divorced from the father of my two children for ten years. I had just started a doctoral program in Clinical Psychology at Pacifica Graduate Institute in Santa Barbara. I felt thrilled with the program and loved the city; I was commuting from my home in Massachusetts to California once each month for a strenuous but delightful three-day weekend of classes, plus working four days a week in a high pressure job as a bilingual psychotherapist, and also throwing pottery on my free weekends.

As our relationship progressed, I struggled with the romance that presented itself in my life, torn by my ambivalence towards men and my fears about committing myself to a new relationship.

September 30, 1990, Journal:

A question in my mind is whether I need the grounding a marriage would give me to do the next stage of my work, or whether it's better for my energy to be alone.

In a way I envy women who do not have men in their lives because they are more free. Yet the Universe has put this Spartan but loving man into my life. Why? The last thing in the world I want is marriage. And that is exactly what my path seems to be calling me to.

Little did I know that cancer was just around the corner. If I had known, I would not have married, feeling that I was somehow "damaged goods"; however, Soli's loving presence in my life came just in time to give me the support I needed to get through the initiation. Studies have shown that married breast cancer patients have a lower death rate than those who are divorced (LeShan, 1977) and that support increases survival time (Goleman & Gurin, 1993). After seventeen months of courtship, Soli and I married, on April 14, 1991.

I suffered tremendous anxiety in the two weeks before the wedding, losing my appetite and six pounds, but then the honeymoon and the moving in together were filled with bliss. After we settled in, we began the task of meshing two different life-styles and dealing with the challenge of bringing a stepparent into the home of a teenage daughter. Soli had never been married before, and my previous twelve years of singleness had given me a long time to get used to having things my own way.

The process brought to a head deep inner conflicts I had about men and brought to consciousness my own sexual woundedness, which began when a close friend tried to rape me when I was nine years old. These wounds needed to be opened up so they could heal. Two months after the wedding, the low-grade depression I had been struggling with off and on for the past twenty years returned with a punch.

June 22, 1991, Journal:

What is this deep inner sadness I feel? Something way down inside of me is crying.

I felt sad [believing] that Soli did not share my passion for learning to do sand tray therapy in my graduate program because of its expense.

My nerves are on edge—door slams bother me.

I long for more *love* in my *heart,* but I feel blocked.

Lord Jesus Christ, what energy pattern do I project into the world that sets up this depression? I will to release it, whatever it is. I don't want to continue.

Financial worry is part of it, but I have enough money for all I need for school, my daughter, and myself. So why worry?

I feel bad when Soli gets upset with me, and I can't keep my boat stable if the ocean is rocking.

I chafe under so much togetherness, people around me all the time, and I have difficulty keeping track of myself. How do I not lose my individuality being with Soli?

The inner force working towards my wholeness wanted to bring to my attention the psychic danger in my conflicting feelings towards my husband. Jung (1974) taught that dreams use the language of symbols to show what is going on inside of a person. "The symbol in the dream has . . . the value of a parable: it does not conceal, it teaches" (p. 32). The following dream soon came:

June 25, 1991, Dream:

I hear rustling in the bed. Soli gets in, saying his bike broke down in Auburn. I am slightly pissed, resenting his interrupting my rest.

Soli is driving me in a motor speedboat on the lake. He calls a dolphin to my attention. I watch and see that no, it is a shark. I pull in the leg I had been dangling overboard, and we zoom towards shore.

Not far from land, there is a commotion with Soli, and he disappears. I am frantic, trying to find my husband's body. I search in an ever-wider circle, but see no trace. I worry about the shark. I go back up on land, and over the bank I see a dolphin zoom in. I go over to ask him to help me find Soli, and here he has brought Soli to me.

Soli is weak and banged up, but conscious. I am very grateful to the dolphin and carry Soli in my arms. I put him on the bed and doctor his wounds.

Same day journal entry:

This dream alerts me to the other side of any ambivalence about Soli: I would really miss him if he were gone.

> Adjusting is difficult, but we've had some real moments of bliss. This month has been hard work, but then sooner or later our complexes would have to bring things to a head. I long to *feel more love* in my heart. Why is it so blocked? What is the shark inside me that bangs up Soli? Or is it external? Is Soli really driving me (on the motorboat)?
>
> Lord Jesus Christ, please help me understand and deal with the man I married. Please help me learn to really love him.
>
> Amen! Selah!

Perhaps the real problem was not just that I could not feel enough love for my husband, but that I could not love the male principle within me enough—the strong part inside of me that would see the world clearly, deeply love my feminine body, and provide for my daily needs. The dream might have been working to heal that inner split. Since Jungian psychology teaches that often the characters in a dream may be parts of the dreamer herself, perhaps the dreamtime Soli represents my inner male.

The first paragraph of the dream sets the stage, showing that I resent his interrupting my sleep and would prefer to stay unconscious, unaware of what is going on around me. This attitude towards my inner male principle could have set up a danger represented by a shark, which I fear has killed Soli.

The beauty of the dream is that it also shows where healing lies. When I ask the dolphin to help me find my husband's body, he not only brings my inner male back to me, but Soli is alive, even though wounded. The energy that I needed to rescue my inner male came from the spirit of the dolphins, those magnificent creatures whose intelligence may actually be superior to ours and who have often in real life befriended and saved the lives of humans (Robbins, 1987). They are not fish, but mammals, which have adapted to life in the ocean, which Jung regarded as a symbol for the unconscious. In this dream the dolphin might represent the ability to take one's intelligence into the world of the unconscious to retrieve lost parts of self. The dream ends with an attitude shift to caring for the inner man.[1]

---

[1]The sections devoted to my understanding of the dreams included in this book are very brief because of the limitations of space. However, each dream a person receives has

Throughout the two years before finding the cancer, my dreams were filled with images of sexual conflict, and I also felt the emotional tug-of-war within that was produced by my ambivalence towards my inner male and towards my husband.

The following journal entry was written four days before I found the lump:

July 8, 1991, Morning

Meditation/Visualization:

I instantaneously burst into tears today when the thought comes to me, "I am married." I feel from deep within that I hadn't wanted to do it. That's why I had so much anxiety around the wedding. Torrents of tears come with deep sorrow.

After much grief, I feel the Lord Jesus Christ saying to me, "You are married to *me*. In marrying Soli, you made a sacrifice of one kind to achieve a greater good."

Later I visualize the Lord Jesus Christ putting his arms around me and saying, "I love you. Open up your heart and let my love in." I try to.

My Prayer :

I totally dedicate my life to my husband Lord Jesus Christ—all parts of me now and forever. I ask Lord Jesus Christ to use me in the way he/she wants and to do the thesis work in the manner he/she intends.

I felt peaceful after this meditation, having made a deeper commitment to allowing control of my life to come not from my ego, which was ambivalent about being married, but from "the Self," symbolized as Lord Jesus Christ, a transpersonal center of wholeness

meaning on many levels, with its richness unfolding the more it is studied. The reader may see many meanings in addition to the highlights I have touched on, or the reader may feel that the dream would mean something entirely different if she dreamed it. No "interpretation" is right or wrong: the goal of looking at dreams is just to get deeper understanding of one's life. Also, the meaning of a dream symbol may get clearer in time, with a person seeing only years later the deepest message that a dream carried.

and completeness, which knew that I needed Soli's physical, emotional, and spiritual support to get through the initiation that was just around the corner. Little did I know when I wrote these words that the thesis work would involve my having cancer!

CHAPTER 3

Finding the Lump

Later in the day on July 8, I got a phone call from my boss at the clinic where I worked, saying there had been a tragedy in the family of my colleague Rebecca, a good friend who was also a therapist on our mental health team.

Her cousin Peter had killed himself over the weekend, and Rebecca was the one who found him dead. I had no idea Peter was suicidal and remembered only the wonderful stories she had told about finding her cousin Peter again two years ago after not knowing him before. She was thrilled by his marriage to Marcela, a beautiful therapist he met in Argentina.

At the wake for Peter, many people who knew him spoke of his tormented history. His mother had committed suicide when he was very young, and he had been tossed about much of his life. Marcela wept profusely and begged, "Forgive him. Just forgive him."

The wake affected me deeply. I worked with suicidal clients every week at the clinic, but the reality of death by suicide never impressed me as strongly as when I was looking at Peter's dead body.

Two days later, on July 10, my daughter Vivien and I were the only two students who showed up for our ballet class, which usually had about seven people. Isabel, our teacher, was upset, and we spent a long time talking with her. She was the youngest of eleven children. Her mother had died from breast cancer when Isabel was about a year old, so she never really knew her. Two years ago, Isabel's older sister Ruth also died of breast cancer. Isabel had been close to her sister and was with her a lot during the time she was ill.

To cheer up Isabel, we made a date to take her to a ballet performance with us. I also offered to get a psychic reading for Isabel from a friend of mine to see whether it could shed any light on the situation with her mother. Isabel was delighted with the idea.

Death was staring me in the face the four days before finding the lump, both from being touched by the story of Isabel losing her mother and her sister to breast cancer and from the direct image of Peter's dead body before me, his life taken by his own hand.

My dreams spoke of an explosion, an apt metaphor for a cancerous tumor erupting and spreading destruction throughout the body, and many images of babies.

July 11, 1991, Dream (Night before I found the lump):

The person I am with looks at something on the porch of my aunt and uncle's old house. Then he says, *"Get ready for an explosion,"* and we try to get out of there as soon as we can.

I have a baby daughter and name her "Myrtle Pearl." I'm getting ready to split suddenly and will leave Mom at the church of my childhood with no way home.

Ever since I started the doctoral program, I had been dreaming that I was pregnant and having a baby, perhaps an image my dreams produced to symbolize the new life that was developing within me. In the dream above, the mother within got left at the church, the place of collective worship. Thus, the dream suggests that to give birth to the gem of new life within me ("Pearl"), I would be splitting from the collective values and ideals of motherhood I had internalized in my childhood.

July 12, 1991, Dream (Day I found the lump):

I am with a woman who is slow: she has a tiny baby and isn't with her. Where does she leave her? In the piano studio. I go in to get her. The baby is screaming in rage. I hold the child and comfort her. I help the mother change her diaper.

A classmate at Pacifica is pregnant! She is wearing maternity clothes. The baby was conceived during last month's Pacifica session. She and her husband had GREAT SEX.

I work somewhere at night and then go to school the next day, going directly there. Someone put the car key into my gas cap. At school, I realize I have no diapers for my baby. I say I'll have to break down and go buy some Pampers.

I interpret the first paragraph of this dream as showing that my inner mother, my capacity to nurture myself, is slow to respond to the needs of the child within me, who is getting pretty mad about not being nurtured.

Next comes the image of new life conceived at school from union of the opposites of male and female, reflecting my excitement over the new world introduced to me through the knowledge gained at school and the thesis work that was gestating.

The closing paragraph suggests that the busy pace of my life, working at night and going to school during the day, leaves me in a situation where I do not have what my own inner child needs for basic comfort: diapers. I was not taking time to relax and "pamper" my physical body by doing things like getting a massage or taking a hot bath to eliminate the buildup of stress from all the activities I had piled into my life.

On July 12, I had a day off from work and was lying in bed awhile after waking up in the morning. I thought about Isabel's stories of losing her mother and sister to breast cancer and decided to do a breast self-exam, a health practice I knew I should do regularly, but only remembered once in a while.

I found a marble-sized lump at the base of my left breast.

My first reaction was disbelief: it was not real. Then a sinking feeling set in, and I knew it did not belong there because it was steel hard, unlike the surrounding softer tissue and the other large globs I often felt. (Now I realize those big globs were ribs.) My emotions froze. I thought the lump was not true and would go away if I felt for it again. But it didn't. Something inside of me sank very low, raising my anxiety sky-high.

I wrestled with what to do for about ten minutes, then called my doctor and made an appointment for 1:00 that same day.

I called Soli at work and told him about the lump and burst out crying on the phone, "I'm so scared!" I thought that cancer equaled death, and if the lump turned out to be malignant, it would likely mean leaving my life, which I had just decided I loved and was absolutely wonderful!

Soli called back half an hour later to suggest we have lunch together by the pond before my appointment.

I went to meet him and melted into the shelter of his embrace, crying. He said he would love me just as much with or without an extra breast and that I could worry for myself, but not to worry for him. I talked about cutting back a day at the clinic because I felt the lump was related to my working too hard, and he agreed to work it out financially.

I felt his love and support. I told Vivien before I left and wept with her too. An overwhelming feeling came into me that I was extremely glad these two people were in my life. I would not have wanted to go through the ordeal of cancer alone. Vivien gave me her copy of cancer survivor Louise Hay's book *You Can Heal Your Life* (1985) and told me I absolutely had to read it.

The woman physician at the clinic examined me and reported, "Well, you do have a lump." But she said not to worry, that lots of lumps like this one showed up during the menstrual cycle and not to think I had breast cancer just because I had a lump.

She checked the results of a routine mammogram that had been taken just two weeks beforehand, and the report was fine. There was no change between the baseline mammogram done a year and a half earlier and this current one. She was casual and did not appear to be worried about me.

She said everybody is always coming in feeling lumps, and if it were still there seven days after my next period, I was to come back. She said that before they marred breast tissue with a biopsy, they needed to see if it were from normal hormonal changes in the menstrual cycle and added, "Don't feel it till then, because it will just make you nervous."

I asked her, "You mean that lumps that are cancer don't feel like this one?"

When she answered, "No, they feel exactly like this one," my heart sank and the fear of death again clouded my mind. But she softened the impact by adding, "But the lump could also be from hormonal swellings."

I was visiting my psychic friend Sara that day anyway and asked her to use her psychic perception to "tune into" the lump. The things she sensed from the lump were that it presented no special danger and that I was not making a choice to die.

I mentally decided that if the lump were not really dangerous and I was not going to die from it, those two factors meant that it must be benign, since I thought cancer meant death. I felt comforted by this state of denial that I had a life-threatening illness.

C H A P T E R 4

Snake Dreams

A powerful force was at work in the unconscious part of my mind even though my conscious mind was denying that I had cancer. The image of the snake came in dreams and synchronicities[1] to bring to consciousness the emotional complexity of the healing process within me.

Although many people fear snakes in today's world, the snake was revered and worshipped in the ancient world. Primitive people believed that snakes did not die as did other animals, but through shedding their skins periodically, emerged reborn into a new life. The Great Goddess of primitive Greek and Egyptian worship was herself originally identified with the serpent, the symbol of renewal and regeneration. The internal Goddess within each person was the Kundalini energy coiled in the pelvis at the base of the spine; through the practice of yoga, this energy could be released to uncoil up the spine through the seven chakras (energy spheres) into the head, bringing infinite wisdom (Walker, 1983).

In ancient Greece, snakes were prominent in the healing temples of Aesculapius, the god of healing, and were esteemed as his sacred servants (Hamilton, 1942).

In the Herakleion Museum in Crete, the cradle of European civilization, is a small Snake Goddess statue from the Temple

---

[1]A synchronicity is a phenomenon Jung noted whereby an outer reality that corresponds to an inner issue or symbol will present itself at a crucial time, as if by coincidence. For example, a person might dream of a certain animal and then see a rare sighting of that animal the very next day. Jung felt that synchronicities lend emphasis to the power of the dream symbols.

Repositories of the palace of Knossos, the legendary home of the Minotaur.[2]

The small faience Snake Goddess

This ancient bare-breasted priestess is shown holding up snakes, because "the ancient Aegean world worshipped primarily women and serpents" (Walker, 1983, p. 903).

In some myths, the Great Mother created a serpent deity, basically a living phallus, for her own sexual pleasure; and in others, she let him take part in her creation of the world. However, when he got arrogant and tried to pretend that he was the only one who made the universe, the Goddess punished him by banishing him to the underworld (Walker, 1983).

In the Pyramid Texts, the heavenly aspect of the serpent was to dispense immortality, and "As the divine phallus in perpetual erection he was the Tree of Life" (Walker, 1983, p. 908). Early Hebrews also worshipped the serpent, and in the Gnostic literature of the early Christian church, the snake in the garden of Eden was praised for

---

[2]In Greek mythology, the sea god Poseidon gave King Minos of Crete a beautiful white bull to sacrifice to him; however, Minos did not have the heart to kill this magnificent creature and decided to keep it for himself instead. To punish Minos for this violation, Poseidon made the wife of King Minos fall madly in love with the bull. She was so consumed by her passion for this bull that she had the architect Daedalus build a female wooden bull that she could sit inside to fool the real bull. She conceived from the bull and gave birth to a son that was half human and half bull, the Minotaur (Hamilton, 1942).

bringing the knowledge of good and evil to humanity, against the will of a god who wanted to keep humans ignorant.

> This view of the Eden myth dated back to Sumero-Babylonian sources that said man was made by the Earth Mother out of mud and placed in the garden "to dress it and to keep it" (Genesis 2:15) for the gods, because the gods were too lazy to do their own farming and wanted slaves to plant, harvest, and give them offerings. The gods agreed that their slaves should never learn the godlike secret of immortality, lest they get above themselves and be ruined for work. (Walker,1983, p. 905)

By eating the fruit the serpent offered her, Eve broke out of the robot "do as you are told" mentality intended by the gods and moved into the capacity for independent thinking. Instead of humans being a race of slave gardeners for the gods, we have become aware of our lives, able to be conscious of what is going on around us because we have the capacity to differentiate between good and evil and therefore are able to make conscious decisions, which can shape our lives in the directions we desire. Thus, the serpent who persuaded Eve to eat the fruit might be seen as a *bringer of consciousness* to humanity instead of a wicked creature tempting Eve to sin. The consciousness that resulted from eating the fruit gave humans the godlike capacity to help form their own lives by being able to choose between good and evil.

New consciousness is new life, and cancer brings new life to a person, either by shedding the old physical body and being born into the life of the next world, or by symbolically shedding old attitudes and stepping into a new, larger life space without discarding the physical body.

Snake symbolism contains the union of the opposites of destruction and rebirth. "Death and growth may be represented by the same tokens . . . by snake symbolism (for the snake appears to die, but only sheds its old skin and appears in a new one)" (Turner, 1987, p. 9). What feels like death may actually be just the shedding of the old skin.

Why then has the snake acquired such a bad reputation? The dictionary definitions of the word "serpent" carry negative connotations:

> **1a:** a noxious creature that creeps, hisses, or stings **b:** SNAKE; *esp:* a large snake **2:** DEVIL **3:** a subtle treacherous malicious person. (Webster, 1965, p. 792)

Perhaps the reason that Webster is so one-sided in his definition of the serpent, seeing it in the traditional view as being nasty and bad, is because along with the snake's urge toward growth and new life comes a powerful force that resists change and tries to pull a person back down into unconsciousness, back into being a slave in the Garden of Eden, doing what one is told without using creative thinking. When the serpentine pull toward higher consciousness is not understood, it may be feared and therefore labeled diabolical by the force of the status quo, which resists change. People do not fear snakes just because they are dangerous. Most snakes are not poisonous to humans and help the ecological balance by eating rodents, which destroy crops. Many more people are killed by automobile accidents than by snakebites, and yet cars are not generally stigmatized as being demonic. Perhaps snake phobia may be connected with the fear of growth and change, the fear of letting the old comfortable emotional skin rupture and die so the body can expand into an emotional skin one size larger.

Like snakes, cancer is also widely feared. All of the dictionary definitions of serpent fit the way people usually think about cancer: noxious, diabolical, treacherous and malicious. Obviously, cancer can kill, but it also offers the person with this illness a chance to shed old skins, old ways of thinking, and step into a larger life space.

Nine months before I found the lump, exactly the length of time a human baby takes to move from conception to birth, a dream directly linked serpent power to the diagnosis of breast cancer:

October 22, 1990, Dream:

Poisonous snakes keep running through the house. I get tired of it and decide to put an end to it. I catch one and flush it down the toilet. As I do, I see it can navigate the septic system. (But I know it can't get out and will die.) I catch another one and am deciding what to do with it. I think of having my daughter get a knife to cut it in two.

Then I have a change of heart and put it into a plastic bag with some bedding material. As I close up the top, I get a telepathic message from the snake that it does not have enough oxygen to survive. I think of punching small holes in the sides. As I am working with the top, I accidentally scratch the snake under the chin, and he loves it.

We begin a relationship. I ask the snake if he is hungry. I ask him what he wants me to do with him, and he suggests going to a zoo. I agree to call some zoos. I'd thought of that too. He is black with pink and yellow racing stripes. He comes from Africa. I ask where, and he shows me the map of the world on his belly. I jokingly ask him if all the snakes of his species come with maps!

I open up an encyclopedia, and a red coral snake gets out. It immediately eats some bugs. I think to check to be sure it is harmless and let it eat up my bugs.

A doctor tells me of a teenage woman patient referring to me. She has evidence of breast cancer but is in such denial, the doctor cannot talk to her about it.

As I analyzed this dream, I noticed that at first the dream Barbara, the dream ego, sees the snake as evil. I do not want to let go of the old to move into the new, seeing only the destructive aspect of the snake. Fearing death, I want to eliminate the snake and try to get rid of it by flushing it down the toilet or by cutting it up. In the dream, I do not get into a good relationship with the positive aspect of the snake, the symbol of the god of healing, until I have a change of heart. Then a personal relationship opens up with the snake, and positive aspects of snakehood emerge, including its eating up pesky bugs.

A similar process happened going through cancer. A deep soul wound was healed, changing my heart and beginning a better relationship with my own healing, opening up my mind into a higher level of understanding, gobbling up some things that had been bugging me.

The dream contains a joke about an ancient image called the "ouroborus," which is a picture of a snake encircling the world and biting its tail. The ouroborus symbolizes nature's cycles of manifesting and reabsorbing, like a tree sprouting leaves, which then die and fall to the ground. These same leaves then decompose and provide nourishment for the tree to grow new leaves. In this dream, instead of the snake's belly going around the world, here the imprint of the world is on the belly of the snake.

The most remarkable aspect of the dream is the final paragraph, where the topic of breast cancer appears immediately after the snake imagery. When I had the dream, because I was in denial, I assumed it meant that a doctor was referring a woman who had breast cancer to me, and I decided to watch for a client with that problem to appear in my office soon. But upon a more careful reading, the dream says the woman is a "patient referring to me." I think the dream is referring to *me:* I already had a malignant tumor growing in my breast at the time of the dream, but perhaps a teenage attitude within me was in such denial that my inner doctor, my inner healer could not talk with me about the cancer yet.

A subsequent dream connected snake energy to the sexual trauma buried in my body from the attempted rape in my childhood. The snake appeared in this dream linked with its phallic energy, first seen in its negative aspect as my fright from the person who molested me waving the snake in my face.

February 13, 1991, Dream:

The friend who molested me is the only one who likes to FRIGHTEN me with snakes. He gets one crawling across the kitchen floor and then is waving it in my face. I hate it.

I go downstairs to go to my bedroom at my childhood house. People have moved the staircase and enlarged the basement. I get into my bed. I see a *huge* snake slither by in the upstairs part. I realize others must be around too. A snake comes out and I grab it, holding its jaws apart. It has fangs going in about five directions. This one seems to have a nail protruding from its mouth, and it strikes me as I am trying to kill it.

I sit there realizing I cannot kill all the snakes. What can I do to get safe? I think of a story and decide to make peace with the snakes. They are important to eat up the bugs and rodents. I want/will them to just stay away from me.

I go upstairs and decide to lock the front door. The latch turns in my hand as I do. Who is there? Dad is on the other side. What a surprise!

Fear is my initial reaction to the image of the serpent in the dream. When I was molested as a child, I was frightened by phallic energy being waved in my face and felt bitten by the experience, marked by the encounter with the powerful phallic serpent god, which I did not understand at the time. In the dream, I first tried to kill the snake; in real life, after the initial encounter, I repressed the memory, trying to kill it by burying it in order to avoid dealing with the painful feelings that accompanied the event. But as my dream attitude towards the phallic energy of the snake softened, the positive side of this force emerged in the image of my father, the phallic energy that united with my mother to conceive my life.

As discovery of the cancer got closer, snake images came more frequently:

May 29, 1991, Dream:

I see a large snake on my plant I am about to water.

In an altercation by the house, a hero is stabbing the lion. No! Stop! Don't hurt him!

He stops after plunging his knife into the lion's buttocks twice. I call for a doctor and a vet saying I have a wounded lion, l-i-o-n, and I am not lyin'!

Both of them come. The lion takes a long piss.

In this dream the snake is connected with the symbol of the lion, which in Christian imagery represents the kingly nature of Christ (Cooper, 1978). Like Christ, the lion in this dream gets wounded, linking the snake image with the paradox of wounding and healing. The Christ nature in me was about to pass through suffering, just as Christ did; however, since the crucifixion was followed by the resurrection, the dream maker keeps a sense of humor about the whole process, putting a joke about lion/lyin' into the dream and ending with the lion relieving himself by taking a long piss!

Only the poisonous aspect of the serpent is mentioned in the following dream.

June 11, 1991, Dream:

I am on a trip with others in a huge motor home with a dysfunctional family.

At the end of the dream, a man is showing us how to spread gas around the yard to kill rattlesnake seeds. The county has vowed to eradicate the rattlers.

The dysfunctional family mentioned here points to the dysfunctional way the different parts of my own mind were relating to each other. My goal-oriented inner male did not listen to the inner child in me, which needed rest and nurture, and the inner mother was in so much pain that she was not doing her job either. A dysfunctional family deals with feelings by repressing them, trying to eradicate them just like the man in the dream wants to eradicate the danger he perceives in the rattlesnakes. The only problem with repressing unpleasant feelings is that they will not stay dead: they pop up in their negative aspects later, like rattlesnake seeds hatching, poisoning the future situation. The dream was trying to bring the dysfunction in my internal attitudes to my attention.

This dream is balanced by another dream showing a more positive relationship to emotions and to instinctual energy, represented by a panther, two nights later:

June 13, 1991, Dream:

I have a pet panther, yellow with brown spots instead of black, who goes with me everywhere. Others at the dump are frightened at first, but then they relax when he listens to me well. I have a snake laid across my forehead.

Since real panthers are solid black and do not have spots, the dream animal seems to be a hybrid of several large members of the cat family. The ocelot, who does have black spots on a yellowish ground, is an animal who hunts at night, sleeping during the day, and whose diet includes snakes. Cats find their food by instinct, and in the dream I have made a relationship with this symbol of powerful instinctual energy by becoming friends with the animal.

The position of the snake on my forehead is on a point that Eastern philosophy teaches is a "third eye." This "inner eye" is thought to be located in the center of the head, just above the top of the eyebrows, in the area of the pituitary and the pineal glands, where the spiritual and the physical realms are thought to meet, giving a person the ability to have intuitive knowledge of things that cannot be seen with the physical eyes (Yatri, 1988). The snake in the dream was lying on this "third eye," perhaps trying to bring it into consciousness.

Ten days later, after awakening and meditating, I had the following lucid dream, which set the theme of sexuality transformation and again includes the image of the snake:

June 23, 1991, Dream:

I am riding my bicycle in unfamiliar territory. I come to a retreat center and know I am supposed to go in there. I go up stairs, up more stairs. A man comes down the last flight of stairs and looks at me with silent knowing. He touches my hand and says to go on upstairs an he will return.

I enter the room. Others are there, and on the floor are mattresses with clothing and underwear on them. I find a place and wait.

The man returns, bringing a baby bull on a rope. He says take it to Barbara. I am frightened, panicked, but there is nowhere else to go.

The bull comes to me and I embrace it, petting it with care and loving it, rubbing its belly and head and back. I feel strong arousal at seeing its penis and desire sex with it. I understand now how the Minotaur got born.

But as I rub and love the bull, it shrinks in size and becomes safer, more feminine—an animal smaller and smaller until it fits into the palm of my left hand. Then it turns into images, metal of three different colors, weapons, helmets, all kinds of objects, all continuing to transform in progression before my very eyes.

I see the value of wearing the three kinds of metal. Briefly it becomes a small black snake, and I shudder, but that too transforms, and then the whole thing disappears, seeming to absorb into my hand.

Now through my right hand, which is raised, I begin to feel a tingle that becomes stronger and stronger, like an electric current. It is accompanied by noise like static, becoming quite loud. The tingle penetrates my whole body and feels like it cleanses me. When it is done, I open my eyes and see a pipe stub in the wall where the energy came out from, even though I had no direct contact with it.

A group gathers around me and says I needed all that. The man steps forth. I was wondering his name, and he whispers into my ear, "John Hay." He holds my hands with great love and tells me he wants to unite sexually with me and talks about how it would not be wrong. I am conflicted, knowing of my vows to be faithful to Soli. I feel the healing effect of the whole experience, energizing!

The man who wants to unite with me in the dream, "John Hay," is not a real person in the outer world, but an inner figure, whose first and last names represent a combination of transformative qualities. In the Bible, John the Baptist announced the coming of Jesus, and John the disciple wrote a Gospel, three Epistles, and the Book of Revelation. So the first name "John" combines symbols of the coming of the divine into my life at a deeper level, being a disciple of the Christ, and writing books. The last name "Hay" is the same as the author of the book by Louise Hay (1985) that my daughter Vivien had been raving about. When I had the dream, I had not yet read the book since I was not as interested in hearing the stories of cancer survivors before diagnosis as I am now. Hay felt that the reason her cancer went into spontaneous remission without surgery, radiation, or chemotherapy was because she learned how to forgive others and to love herself more deeply.[3] So the name "John Hay" combines the previous qualities of the name John with healing from cancer and learning to love and forgive oneself.

Transformation was calling to me, through the metaphor of sexual union with "John Hay" in the dream. Transformation was trying to get inside of me, but I hesitated, fearing that uniting with this energy might threaten my stability in the outer world, represented by my marriage to Soli. Yet, I felt the healing and energizing effect of the transformation I was headed for. The dream gave me a taste of the

---

[3]Hay's healing may have worked through a process similar to the way the Christian Science church channels healing energy by accessing the spiritual level through prayer to focus mental power into healing the body. However, *A Course in Miracles* (1976) teaches that if one is afraid to heal by the power of spirit alone, a patient should also use modern medicine. I knew I did not have enough faith to treat the cancer just with prayer. I wanted doctors too! I feel strongly that a person diagnosed with cancer should work together with a physician to find and use the best medical treatment possible—plus use the healing power of love and forgiveness to mobilize the immune system and boost the effectiveness of the medical interventions.

outcome before going through the fire, giving me nourishment to sustain me through the perils of the journey.

This dream resonates with the transformative aspect of the snake, an important mythic symbol in practically every culture. Campbell (1968) related the story of how the snake got its power from the pre-biblical tale of Gilgamesh, a legendary Sumerian king. Gilgamesh set out to find the "watercress of immortality, the plant 'Never Grow Old'" (p. 185). After many adventures and trials the hero finally dove down into the bottom of the cosmic sea and plucked the plant. However, he was so tired after the dive that he lay down to rest before eating the plant. "But while he slept, a serpent smelled the wonderful perfume of the plant, darted forth, and carried it away. Eating it, the snake immediately gained the power of sloughing its skin, and so renewed its youth" (p. 187).

The following snake dream links the cancer with themes of sexuality and trauma to the physical body.

July 7, 1991, Dream:

My girlfriend and I are going up for class by the lake where I grew up. She wants to go wading through the water. I go on shore. Then she is sitting on the wall sunning herself. I comment on how suggestive her pose is, and she says she doesn't care. The only thing she has on is a bikini top. Her legs are crossed, exposing her beautiful blonde pubic hair in an invitational style.

A woman has two bodies and two operations. The one who got the tubal was fine, but the one who got the appendectomy was writhing in pain.

At a mountain home, by the porch are a bunch of huge poisonous snakes. I get one that looks dead and am checking it out. It still has a pink mouth. It trembles a bit when it warms up but does not go into action.

My former husband is outside dealing with the snakes. They come at him, and he tries to grab them but keeps missing. Then the big rats come out, cute and friendly critters. Don't hurt them! Finally he puts the snakes behind the car and rocks the car back and forth over them, to squish them up.

A man comes to the door with a knife to kill the young girl that lives there. He says he will slit our throats while nobody else is there to keep us from stopping him.

I say to him, "Don't be silly. We are all friends here. You don't need that knife," and I break the blade in half. He reconsiders and becomes friendly. In the end he is a boy and passes by the girl he formerly wanted to kill.

The opening paragraph could represent a shadow aspect of my personality,[4] a free and open sexuality that had been repressed in conscious life because of my puritanical Protestant upbringing and my own sexual trauma from being molested as a child.

The two surgeries mentioned in the second paragraph of the dream correspond to the only two operations I had undergone in my life at the time of the dream: a tubal ligation when I was twenty-seven and an appendectomy when I was sixteen years old. The dream suggests that the pain of the appendectomy at age sixteen was still having a negative emotional impact on my body.

Some massage therapists who do deep bodywork say that feelings get embedded in the physical structure of the body, "the issue in the tissue," as if we had a separate body for our feelings, which attaches to and interpenetrates the physical body. When I was sixteen, my appendix had ruptured following several days of the most severe physical pain I had ever known. Even though the physical body had recuperated from the exploratory surgery, which had discovered my ruptured appendix had spread all over my abdominal cavity, the emotional trauma from all the pain involved had not healed. According to the dream, my emotional body was still wounded, writhing in pain from the appendectomy.

Snakes follow in direct sequence to the pain in the dream. During the week I had stayed in the hospital after the surgery at age sixteen, the woman in the bed next to mine was dying of cancer, and that contact put the fear of death from cancer into my mind.

When I read these words of Kahlil Gibran (1966) the following year, I realized the agony of that hospital bed had changed my world:

---

[4]"The shadow personifies everything that the subject refuses to acknowledge about himself and yet is always thrusting itself upon him directly or indirectly—for instance, inferior traits of character and other incompatible tendencies" (Jung, 1968, pp. 284–285).

> Your pain is the breaking of the shell that encloses your understanding.
>
> Even as the stone of the fruit must break, that its heart may stand in the sun, so must you know pain. . . .
>
> It is the bitter potion by which the physician within you heals your sick self. (p. 58)

Going through the appendectomy at age 16 marked a change in my outlook, a transition from the blissful "death cannot touch me" innocence of childhood into young adulthood as I peeked at mortality for the first time, realizing that without medical intervention, I would have died. I felt changed by that illness, as if its pain and suffering had hollowed out more room inside of me to fill with life. Anyone who has ever survived a serious illness can attest to the transformative power that a brush with death can have to broaden one's horizons and give a new perspective on health and life itself, shedding the old skin and stepping into a larger one.

I grew in wisdom from the surgery, but maybe its trauma to my body never healed. Perhaps I needed some method to heal from the physical effects of the damage done to my body by the surgeon's knife in the process of saving my life. The physical shock of all that pain went underground, and repressed trauma does not magically disappear. It festers.

When the snakes in the dream were squashed, the inner male became violently aggressive toward the young feminine within. I had squashed a lot of my sexuality because of taking on the puritanical sexual values of my church; I had squashed the memory of the physical pain of the operation, and I was squashing the message from my dream world that I had cancer (the dream of October 22, 1990, where my inner doctor could not talk to me about the cancer because I was in denial). This former attitude my inner male held of trying to repress my feelings, squashing the snakes up under the car wheels, was destructive to the young, innocent, feminine side of my personality. Since the figure who has cancer in the former dream is a teenager, perhaps this dream sequence hints that the cancer might have been seeded by the unresolved emotional trauma from the pain of the appendectomy at age sixteen.

When the destructive energy of this inner male figure is directly confronted in the dream instead of trying to repress it, that same energy becomes friendly, like the man at the end of the dream. Instead of trying to cut off from the painful parts, in the dream I break the knife, breaking my pattern of separating from my own pain and fear.

The dates of these dreams show the psychic buildup leading to the discovery of the lump on July 12, 1991. On a conscious level, I had been denying that I had cancer, but for the previous year, my dreams had been sending me messages in symbolic language to try to bring to consciousness the illness and the transformation that was about to come into my life.

The cancer in my body had probably been growing somewhere between five and twenty years, since the average cancer cell doubles every three months; a single cancer cell needs an average of ten years to grow into a lump big enough to be felt, a billion cells, one centimeter in size (Love, 1990). My 1.5 cm. lump probably only reached detectable size the month before I found it, and during this month, I dreamed about snakes four times.

Early detection increases survival rate, because cancer is a battle between the destructive force of the illness and the life force of the body. If breast cancer is stopped early, while the life force of the woman is still greater than the force of the illness, the fight is easier to win; however, when the malignant cells that spread to other parts of the body have more time to get established and begin to affect vital organs, the body has a bigger enemy to fight and less strength to fight with, so the battle is more difficult. Even so, as long as one is still breathing, hope remains: spontaneous remissions have been known to occur with even the worst kind of cancer, Stage IV, with metastases to distant parts of the body.

Like many women, I was not in the habit of doing regular breast self-examinations. I denied the danger of cancer because I was in good health and had none of the known risk factors. I felt too busy to bother with the self-exam and could not visualize myself being a cancer victim; cancer was something I thought only old people got, and at age 42, I felt that neither death nor cancer could touch me.

CHAPTER 5

Diagnosis

Several days after finding this strange mass in my breast, I meditated with my crystals, putting my amethyst crystal on the lump and visualizing it dissolving. [Note: I had been using this amethyst crystal to treat minor ailments such as joint pain and stomachache. However, crystals are *not* recommended for treating cancer because their focused energy might tend to increase the vigor of the cancer itself.] Even though I was denying that I had cancer by telling myself that the lump was only a cyst, I really wanted to get rid of it. The next day, I disregarded the doctor's suggestion to wait until after my period to feel for the lump again and checked it. My stomach sank as I felt the steel-hard foreign mass, still there, but I clung to the hope that my period might dissolve it.

During this month of waiting, I was meditating one night and wanted to ask Lord Jesus Christ, "What is the meaning in my life of this lump?" However, my mind could not get the word "lump" into the thought forming in my brain, and the question that came out was, "What is the meaning of this CANCER?" I mentally struck out the word cancer and again tried to ask the question, "What is the meaning of this lump?" But again, the question formed with the word cancer instead of the word lump. I tried to ask the question a third time, and the horrible cancer word kept inserting itself into my thought.

The answer to the question I kept trying not to ask was that it was about slowing down the frantic pace of my life; however, instead of taking the message to heart, I stopped meditating because I did not like hearing the cancer word and thought I was just thinking negatively. I mentally went right back to my denial that I had cancer.

A week after finding the lump, I had the following nightmare:

July 18, 1991, Dream:

I hear the big bang explosion. I think, "Who have I lost, Mom or Dad?" I do not want to lose either one, but think it is Dad. Then it is Mom, and she is dying. The gas had not been coming out of the system properly, and Mom was messing with the pipes and valves to try to make it work. The leak produced exploded when it reached an open flame.

I finish up what I am doing and go to where she is stretched out on the ground. Others are there. Dad is caressing her torso and expressing his love. I go to tell her of my love for her and my presence there. She says, "*My cold feet,*" complaining almost hysterically. I think she is dying from the bottom up, like Socrates. I want to hold her feet.

I am in a car with my younger sister, who is driving backwards. I watch for cars and call that some are coming. She doesn't respond, and I shout "Take care!" so loud that I actually say the words out loud and wake up Soli.

This dream gives a vivid image of the part within me that was dying to make room for the new life that wanted to come, like the acorn that must die to sprout into an oak tree. In the dream, the person who represents my image of being a mother dies in the explosion produced by an energy leak. The internal script of mothering I had been reading to myself said that to be a good mother, I should take care of everybody (*except myself*). In this pattern of nurturing, I had been giving out more energy than I took in, depleting my own energy supply like a gas leak, and the results were disastrous.

The dream repeats the image of an explosion used one week earlier in the July 11 dream and goes one step further by showing the source of the energy leak that produced this explosion.

In the dream, my internal mother is lying on the ground, hinting that she is beginning to get grounded. She is afraid of the death stage of the initiation she is going through and complains hysterically about her cold feet. Anger is a hot emotion, but fear is a cold one; when afraid, a person may "freeze up." My response in the dream is to go hold those feet of hers, which are so cold, trying to warm the part of the body that is in closest connection to the earth, attempting to further ground her.

The last paragraph shows that trying to hang on to my traditional image of mothering in this way heads me backwards, and the message that I needed to *take care* in a different way came out so loud and clear that I woke up Soli. I needed to learn to add my name to my list of people who needed caring for so I could replenish my own energy supply, preventing an energy "gas leak," which could cause another explosion.

Two days later we took our friend Isabel to a fantastic ballet at a theater called "Jacob's Pillow" in the Berkshire Mountains. On the trip up, Isabel said the psychic reading she got from my friend had been very helpful. The reading said her mother had found out she had breast cancer while she was pregnant with Isabel, and she had such love for her unborn child that she refused to have the abortion her father wanted her to get to save her own life. She had felt a strong soul connection to Isabel and wanted to give her the gift of life. Isabel had never before thought about the possibility of her father wanting to abort her, but then she remembered that when her sister Ruth was pregnant and found out she too had breast cancer, her father had pushed Ruth to get an abortion.

Ruth made the same choice her mother did, of refusing an abortion and thus delaying chemotherapy until after the baby was born, sacrificing her life for that of her unborn child.

I knew I had a lump in my breast, but I did not mention it to Isabel. I listened intently to the progression of the illness in these two women, and I wondered what I would do if I were in Isabel's mother's position. I asked Isabel what she would do in her mother's place, and she said she would keep the baby. Isabel was healed by knowing that her mother really loved her and cared for her and that her mother's greatest sorrow had been not being able to care for this baby she loved so much.

My period had been due July 19, and by July 22 it still had not arrived. My cycle was very regular and was never that late. Even though I had undergone a tubal ligation many years earlier, I had been dreaming that I was pregnant for the past year and a half. I thought to myself, "Shit! Maybe I have breast cancer and I am pregnant, just like Isabel's Mom and her sister, who both died that way."

Finally, the following day, on my airplane flight to my classes in California, my period came, four days late. I felt enormous relief. At least if I had cancer, I would not also have to make the terrible choice of sacrificing either my life or the new life within me.

> *July 28, 1991, Meditation:*
>
> *I visualize my lump, and it is the face of a small monkey, reddish-colored hair, hissing and ANGRY! I feel it is involved with my need to speak out my anger. I wonder what I am angry about.*

I could not bear the suspense of waiting the prescribed "seven days after my period" to see whether the lump had disappeared, so on July 28, two days early, I checked again. The breast mass was still very much there and frightened me because it felt strange, as if it might be getting bigger and might have a different shape, as if it were funneling down into my breast.

I got out of bed, and my roommate asked if anything were wrong. As I was telling her about the lump, I suddenly broke into uncontrollable sobs. I could barely pull myself together enough to eat breakfast, but hunger eventually won out over despair.

I called Soli, weeping and feeling as if I wanted to go home RIGHT THEN and just be in his arms. I asked him to make a doctor's appointment for me as soon as I got back, and he did. I tried to hold on to a faint glimmer of hope that two days later the lump might be gone, but inside, I knew that it would still be there.

Tears kept coming to my eyes during that whole day of classes, and friends held my hand in gentle support. I talked with lots of my women classmates about lumps and got more information.

I asked my friend Ellie to tell me the whole story of how she had survived her breast cancer five years earlier. She said she had been very busy with school and had little time to attend to the lump in her breast. Her doctor told her in November that it was just a cyst, but the following July she had a dream that clearly said, "Ellie, you have breast cancer." She went back to her doctor, and he did not believe her; however, to humor her, he sent her in for a mammogram. When it came back negative, he said, "See, I told you it was nothing." Because of her dream, she was not satisfied and went to a surgeon.

He tried to aspirate the lump, but could not get anything. Two days later she had surgery to biopsy the lump, and the lab report came back malignant. She knew something was wrong when they said to call her husband.

Another friend whose mother had died of breast cancer showed me the scar she had from a benign lump, which was recently taken out. I was comforted by seeing how neatly the incision was done, following the line of the areola. She recommended I definitely have my lump removed. She told me the story of a doctor's wife whose lump biopsy results were misread at the lab as benign, and the woman died. She said, "Breast cancer is a killer." Terror struck inside me as she talked. I thought, "One cannot trust mammograms or biopsy reports, and women are dying all over the place from breast cancer!"

At noon I went to the beach near the school, and I saw a sea gull dying, lifting its head as if it wanted to cry out but could not. A black cat jumped out in front of my roommate and me as we drove back to the dorm that night, and it scared me. Were these events synchronicities? Omens? The bird dying on the beach was very sad. I asked myself what was dying inside of me and willed to let go of the old, whatever it was.

I dreamed of an explosion again, but this time without the feeling of helplessness:

July 29, 1991, Dream:

I have a weapon, a gun that discharges a shot with a beautiful firework pattern from a staple that interacts with a coil of wire.

I see a blue blob with a gallon metal jug of fuel, which is breathing a low fire inside one of the machines at the Laundromat. I go to report it to the man running it—danger of explosion!

I start a weekly women's group of healthy persons meeting in my home. I have company and am feeding people.

Perhaps the willingness to let go of my old way of life, like the snake that sheds its old skin, set the stage to have a weapon, a means of defense, and to be able to see when danger was brewing. The ending of the dream shows I am beginning to nourish unity in the healthy feminine inner parts of me.

After this dream, I noticed a new attitude forming inside of me and heard myself saying to the people I talked with, "I feel I am strong enough to survive cancer, strong enough to tolerate chemotherapy."

On August 1, the day after I got back home, I noticed myself repeating this statement to my supervisor at the clinic where I worked. I happened to be consulting with one of the physicians on a mutual client that day, and he asked how I was. I said that married life was great, but I was concerned about a lump in my breast. He said that cancerous lumps are hard, very solid, and he offered to feel my lump. At first I felt confused and imagined he was making a pass at me, but then I realized he was being professional and was genuinely concerned for me as a friend and did not want a cancerous lump to go undetected.

I told him I would take him up on his offer if I felt I needed a second opinion after seeing my regular physician later that day. As I left, he asked me if I prayed. I said, "Yes! And I am doing a lot of it at present!"

My primary care physician looked quite concerned when he examined the lump in my breast. He set up an ultrasound for the following day and an appointment with the breast clinic for the following week. He said that lumps are tricky, and some hard ones are not cancer and some softer ones are. He talked about the possibility of my lump being a fibroadenoma, a noncancerous mass of glandular tissue. I liked the sound of "fibroadenoma" and decided that was probably what it was.

The next day I took time off from work around noon to get an ultrasound picture of the lump. The nearly deserted ultrasound clinic had an eerie feeling, and the gel used with the machine was cold on my bare skin. I felt sick when I saw a solid-looking black mass show up on the screen. I thought it said 5 cm beside it. I telephoned my physician that afternoon to get a verbal report of the ultrasound results, and he sounded relieved. He said not to worry, that the mass did not have many of the signs of cancer, but it also did not have all the signs of a cyst. He said their best guess was that it might be a cyst that had bled into itself, which could explain how it popped up so suddenly. He knew that no abnormalities had shown up on the routine mammogram I had had on June 26 or in the routine Pap smear and breast exam I had had on March 20.

I felt partially comforted. Two days later, I had the following dream:

August 4, 1991, Dream:

I get two small mouse-hamster critters in a cup with a slice of mandarin orange. I bring them home and want to air them to be sure they will survive. They jump out and run off. My dog chases one and catches it. I jump on the dog and open her teeth, taking out the creature. To my sorrow, it is not moving, and blood is coming out. It is frozen in a fierce pose, this tiny creature trying to defend itself against the dog hundreds of times bigger. I feel very sad.

At church, I get separated from Soli when I go to brush my teeth. Then a colleague at work is there with some friends, and he is in bad shape: high on drugs and suicidal. They ask me to get some help. I want to take him in myself but don't have a car. Nobody else there helps. I feel helpless and responsible to get him to services.

I see pea-size lumps all over my sides. At first I want to pop them like pimples, but then I think they might be cancers and popping them might open them up to spread.

The initial image in this dream is helplessness in the face of an aggressor much larger than oneself; at the time I did not connect that feeling to the fright of facing cancer and wondered where I felt that same kind of terror in my life. The only connection I could find was the feeling I got when confronting the monumental stressors of the clients I worked with at the clinic, where my clients were daily faced with racial discrimination, crime, poverty, substance abuse, and AIDS.

In this dream, the developing inner male was high and suicidal, and I felt helpless and responsible, two of the keywords the Simontons (1978) found predominant in the emotional profile of their cancer patients. Even though I consciously denied feeling helpless, this emotion comes out clearly in the dream and is directly linked in the final paragraph to the possibility of cancer.

On August 7, I had an examination by Dr. Orr, the breast clinic oncologist surgeon. He said that he had felt only one other lump like mine that turned out to be malignant, and he had felt hundreds just like mine that were not, but he did a needle biopsy to be sure, drawing out some contents of the lump with a needle to send to the lab. He said the needle biopsy would not hurt much, that it was similar to drawing blood, but he was wrong. It hurt twice as much as drawing blood! The needle biopsy took about a minute, and I was glad when it was over. Dr. Orr said that even though he was pretty sure this lump was not cancer, just to be on the safe side, the lump should come out. I agreed totally, and he scheduled the surgery for the following week. My heart raced as I realized that my breast would come under the knife. He said that because the lump fell so low in the breast, the scar of the surgery would be hidden in the folds of my breast. I told him, "I'm not worried about the scar; I just want to STAY ALIVE!"

Initially I felt relief that I could finally get this alien thing out of me. The nurses took me down and gave me five appointments: one with the lab to get bloodwork and a urine sample, one with the education nurse, one for a physical exam, one with the anesthesiologist at the hospital, and one for the surgery itself on August 15. I felt slightly overwhelmed at all the commotion that surgery involved. I wanted to avoid general anesthesia if at all possible, because the thought of going under felt a bit like death, and I promised myself I would have the operation with just local anesthesia.

The next day Vivien called me at work to say that Flower (her five-month-old pet guinea pig that Soli and I had given her for a present at our wedding) was acting strange. Flower was having difficulty breathing and was not moving much. When we got home at 6:00, Flower had already died, and Vivien had buried her in the back yard. I wondered why such a beautiful, healthy young creature was taken, especially when we all loved her so much, and she herself was so loving. We all felt very sad. I wonder whether Flower had cancer, which can also strike a seemingly healthy person and rob that person of her life.

Soli, Vivien, and I had been going to family therapy occasionally to help us adjust to our new family arrangement. At family therapy that evening, we talked about our sense of loss at Flower's death. Soli mentioned "minor physical flaws in us," referring to the rash on his hand and the lump in my breast. When the therapist responded, "I sense a great deal of concern in you, Barbara, around this lump," I realized that wondering whether or not I had cancer weighed very heavily on my emotional state.

The next three days, from August 9 through 11, Soli, Vivien, and I joined a group of 17 other hikers for a three-day backpacking trip in the White Mountains. The first day was beautiful and inspirational. Rain fell all that night and all the next day, making the seven-mile trek to the second hut quite dangerous, hiking in ankle deep water most of the way. I was battling hypothermia, bursitis in my knees, and fear for the safety of Vivien, who took off with others ahead of us. The last mile was straight down very steep rocky cliffs, and the only thing that kept me going was fear for my safety. The lowest point of the whole ordeal was when I had to take a pee in the woods, and I was so weak and tired that I accidentally urinated right on my hiking boot. I was really pissed!

I felt tremendous relief when at last the destination cabin came into sight six hours after we left, and we found Vivien safe inside, having arrived about 45 minutes before Soli and me. The physical challenge of survival in this wilderness area so absorbed my mind that for the whole day the thought of the lump in my breast did not cross my mind even once!

But my alarm came back when I returned home and saw the educational nurse at the breast clinic. I was impressed by the elegance of this beautiful woman with long red hair but puzzled by her speaking to me in the concerned tone a person uses when talking to someone who might be in danger of dying. She said that four out of five lumps turned out to be benign. My coping mechanism of denial had set back in, and so I thought, "I only have a cyst, not cancer, so why is she looking at me like I might be seriously ill?" My panic rose when she said they would definitely put me under anesthesia for the operation. I told her I wanted it done with local anesthesia only, and she said they could try that method.

During the meeting of our team of therapists that day, my colleague Rebecca took us over to her office to show us the sand tray done by a teenager who had suffered brain damage as a child. We watched while she took two flash pictures of the tray to remember it. In the disorganized sand tray picture, two babies were lying face down in the sand. Rebecca said, "Let's see what is buried under this mound," and she brought up another baby who had been completely buried.

As our director brushed off this little creature, all seven of us in the room saw a flash of light. Two therapists said they heard a click preceding it coming from the direction of Rebecca's camera, which was lying untouched behind her. Rebecca said it could not possibly have come from the camera since the flash button was closed down.

We checked the lamp near the camera to see whether a bulb could have burned out, but the lamp was not plugged in. Rebecca said, "There is too much energy in this room!"

We all went back into our meeting place to talk about the strange flash of light, which seemed to come out of nowhere. Rebecca said that she had attended a conference on near-death experiences the past weekend, and many people reported a click and a flash of light before their near-death encounters. I suggested that maybe the client had a near-death experience, and Rebecca said that my theory might well be true, judging from what she knew of his early childhood history.

Rebecca then talked about the out-of-body experience she had when she was in a near fatal car accident at age sixteen. Her liver was damaged; she had extensive internal injuries, and the doctors expected her to die. After the accident she had seen her body from outside of it and had an experience of Light, Love, and a heightened sense of Life. She recovered fully within five months.

Rebecca then talked about the way a native healer, called a shaman, learns about healing. Instead of studying books on anatomy and physiology, the shaman gets sick or is wounded and comes very close to the point of death. If she can survive the fire of the illness, the shaman then has direct knowledge of how healing works from the personal experience of being healed. A shaman is therefore sometimes called a "wounded healer" (Halifax, 1982).

Rebecca had always felt embarrassed by the unsightly scar resulting from her near death "accident." The previous year, she had tried plastic surgery to restructure it. However, the scar remained basically the same as before, and now she regards it as her shamanic mark. She reframes the scar as marking her own important near-death experience, feeling that perhaps this "initiation" was the beginning point of her career in the healing profession of psychotherapy.

Later that afternoon, a teenage client of mine formed a beautiful picture in the sand tray in my office from the miniature objects available to use in this form of art therapy. Unknown to me at the time, the picture she made mirrored my own inner situation.

The sand tray shown had a man working at the bottom, houses around the edges, a pond with turtles and canoes, several young girls doing athletic things, and an ambulance taking a sick person to the hospital. The building at the center of the right side was on fire, and two fire trucks were headed toward it, with all of the people safely

Client's Sand Tray, August 13, 1991

getting out of the house. In the upper right hand corner was a church wedding, with two priests and the couple's family all around.

At the very end, the client put a 15-inch rubber rattlesnake into the center of the picture heading right for the wedding, burying the middle part of the snake's body, leaving just its head and tail exposed. She said the bridal couple and the people at the church did not know the snake was coming. In the sand tray, an army jeep had crashed on the snake's tail. I wondered what the snake was all about. This particular client was very quick to catch on to everything that was going on around her, and her sand tray was both about her own life and at the same time about my inner situation, which she had intuitively picked up. Soli and I were like the bridal couple who did not know that the cancer snake was headed right our way.[1]

On August 14, I went to the hospital to talk with the anesthesiologist about surgery the following day. The young attendant who came to take me to the right room asked if I wanted to go in a wheelchair. I was jolted by the thought of being so sick I would need a wheelchair to get down the hospital corridor, especially since I had just been up and down mountains for three days with a 15-pound pack on my back. I refused the wheelchair with an emphatic "No!"

---

[1]Later, after I had recovered from the cancer, when I asked this client about why she had made this sand tray, she said she *knew* that something was going to happen to me, but she just did not know exactly what it would be.

I was very clear with the anesthesiologist that I wanted the operation with a local anesthetic only, and he said he would try. He had me sign a consent form that said I realized there were all kinds of possible dangers involved with anesthesia, including horrible things like paralysis and damage of other kinds.

That afternoon while I was in session with a teenage girl who was talking about fighting her suicidal urges, the phone rang, and my secretary asked if I wanted to take a call from Dr. Orr. I said certainly, wondering what he could be calling me about. Dr. Orr asked what time I got off work and whether I could see him that afternoon at 5:30, leaving my work a bit early, to talk about treatment options.

I said yes.

I asked if there were cause for concern, and he said yes, there was. I asked if I should bring Soli along, and he again replied yes. I hung up the phone and burst into tears in front of my client.

She asked if I wanted to be alone, and I said no. I said that the reason I had already told her that I would not be in the clinic on the following day was because I had to have a little operation, and I was worried that something might be wrong with me. She said with a broad, radiant smile, the most beautiful smile I have ever seen her give, that things would be all right, she was sure. She said not to worry.

After I finished the session with that client, I dried my tears and headed for the bathroom to cry a few minutes before my next client, but my following client was waiting for me on the couch by the bathroom.

During her session, this woman talked about her last surgery to remove a tumor, without my mentioning anything about what was going on with me. She said, "Suppose I am going in for surgery tomorrow—I go to the hospital today to talk about it. Then, you know what they do when they take you in for surgery the next day?" She said the Spanish words for eating up and made motions with her hands as if eating chunks out of a fish. She said she would not go in for another surgery, that if she had a recurrence of her tumor, she would just die from it. These words did not comfort me! Without my saying a word, this woman had unconsciously tuned in to my apprehensions about surgery the following day and verbalized them, bringing my fears out into the open so I could get a good look at them.

I caught Rebecca and my supervisor as they were about to leave and told them something was wrong. Soli came in just then, and we all cried together. Rebecca offered to be present during the surgery, and I said thanks, but it would not be necessary.

Shaken up, Soli and I went to talk with Dr. Orr. He said, "Well, I guess you have figured it out by now. The cytology from the needle aspiration came back malignant."

CHAPTER 6

Change of Heart

Dr. Orr explained three possible treatment options:

1. Mastectomy, followed by chemotherapy: good cure rate, but not his recommendation. The cure rate with lumpectomy and radiation treatment is about the same as mastectomy.
2. Lumpectomy, chemotherapy, then radiation.
3. Chemotherapy, lumpectomy, then radiation.

My first question when he said I would need chemotherapy was, "Will I lose my hair?" I cringed when he said I probably would, but I could wear a wig.

He recommended the third option since the real trouble was not the lump itself, but the messengers it sent out to other parts of the body. Surgery weakens the immune system, so he said having the chemotherapy first made better theoretical sense. However, the only way he could do option 3 would be if I joined the study he was doing to randomly try out options 2 and 3. If I decided to enter the study, he would put my name into the computer, and I would randomly be assigned chemo first or surgery first. Getting the results of the computer decision would take about ten days. He said that if it were his wife, he would do the chemo first.

Soli and I talked it over and decided to have the lumpectomy the next day. I could not bear the tension of waiting ten days knowing that cancer might be spreading itself around in my body.

Dr. Orr recommended a biopsy of the lymph nodes to see how far the cancer had spread, so I signed a new consent for "Lumpectomy and Lymph Node Dissection." He said sometimes patients have some numbness in the arm after taking out the lymph nodes, but usually the numbness eventually goes away.

I told him about meditating and getting "cancer" instead of "lump" in my question and said that something within me had known it was cancer. He said he believed that. He assured me that my chances for survival were very good, adding that he could intuit by a person's attitude which women would make it through breast cancer and which ones would not. He said the "whiny ones" did not do as well and emphasized that breast cancer is a treatable illness; we just had to decide how to approach it.

Dr. Orr's forthright attitude and confidence that I could get through the cancer started to change my mindset from one of fear to one of hope and courage, attitudes which contribute to heightened immune system functioning. I told him that shamans in every culture are the persons who have had a major illness and have come back from the experience able to facilitate healing in others. When I told him I thought this experience in my life was a shamanic initiation, Dr. Orr said, "Well, I never heard that one before!"

When my husband and I got home that night, we took a swim in the lake by our house. I enjoyed the movement of the water over my strong, perfect, muscular body and could hardly believe cancer was growing inside, still thinking that maybe the cytology report was a mistake. I saw my neighbors but did not tell them I had breast cancer.

After the swim, the first person I called was my psychic friend Sara. I told her the lump was malignant and asked her to "tune in" to what was going on. She immediately did a reading and called me back in half an hour, saying that I had created this illness in myself because of needing to learn how to receive healing, love and joy into myself. She said it was not about healing others; it was about receiving healing into myself and that I was in control of the illness, being free to choose to die, although she did not see that choice. She saw me fighting. She saw hot flashes in my armpits and said that the cancer was already receding from my fighting it. She said this illness would turn my life around and open up wonderful new possibilities of *being* that I had never before even dreamed of, with more joy!

Vivien was visiting my parents, and I called the three of them next and told them the lump was malignant and that I had cancer. I asked them to pray for me.

Next I called my son Robin, who lived with his father. My former husband answered the phone, and I asked if he had a few minutes to speak with me. I told him I had breast cancer and asked if we had any unfinished business left over between us from the divorce. I burst out crying and said that with all that went on in the divorce, I never really had a chance to tell him how much I loved him. I asked, "Are you still mad at me for anything?" He said he was not angry anymore and that we had some bad times, but we had lots of good times too, and those are the things that he had been remembering lately.

He also said he really appreciated what Soli was doing by sending our son Robin to China that fall for a semester abroad. He agreed with me that Soli is a good man. I asked him to pray for me, and he said he would pray real hard. His own mother had been diagnosed with breast cancer five years ago and had just died four months earlier from cancer in another part of her body. My former husband and I agreed to be friends again, and I felt joyful and healed by the interaction. Robin was not home at the time, but he had him call as soon as he returned.

I called my older sister, who lived three thousand miles away. When I told her this experience was about learning to receive love and healing, she cried and said that message was for her too, and we agreed that we are two peas in a pod. She said that even though she meditated and prayed and ate healthy and everything, still, something in her soul just HURT. I knew exactly what she was talking about. We prayed together on the phone, and she asked "Mother-Father Goddess" to surround me in healing light and cast out the darkness within me. We talked about how all the old wounds were now forgiven, and I told her that things with my former husband were forgiven too, and she wept for joy. I felt her love surround me.

Robin called and we talked. He was supportive and affirmed my feeling that spiritually we are always near to each other. I told him I would break his neck if he did not write to me from China, and he laughed and agreed. He said he would pray for me.

My supervisor called to check on me, and I told her the news. I called Rebecca and took her up on her offer to be present during the surgery. I decided to rally all the love and support for myself that I could find.

Soli came upstairs, upset. He was feeling left out of my group of supporters and wanted to be alone with me to process his feelings and did not want me to keep talking to other people on the phone. I decided that if this lesson were about learning to get what I needed for healing instead of attending to the needs of others, I had better start right then. I told him I was sorry, but I needed all the support I could get for surgery the next day and continued talking on the phone. I called a few more people, including our mutual friend Marie who is an osteopath; she was shocked and concerned, and I appreciated her talking with Soli too for awhile.

We finally got into bed about 11:00 P.M., and Soli was exhausted. He talked about his sadness and wanting to be alone with me to cry together, but I could not get to tears just then. I wanted to make love, but I saw how depressed he was. After we talked a bit, we held each other and made love. As we both reached climax, I burst out crying from the bottom of my heart, and we both let our sorrow flow and cried loud and hard together.

I showered and then prayed, connecting with Lord Jesus Christ and asking for strength to go through whatever was ahead of me. I again turned my life over to Mother-Father God. I started to get distracted by thoughts of the surgery, when in my mind I heard Lord Jesus Christ say, "I now heal your soul." With my eyes closed, I saw a mental picture of Jesus, dressed in a long flowing purple robe, touching my heart center. At the same time, I felt the physical sensation of touch right over my heart. This healing touch was very light, yet in that one instant, a deep wound sealed up in my chest area, a wound that seemed as if it had been aching since before my birth. I experienced a marvelous taste of wholeness, bringing tears to my eyes and flooding my soul with pure joy. From deep within, I intuitively knew that since my soul was healed, my body would follow suit, and I would survive the cancer.

C H A P T E R 7

Surgery

I did not sleep much during the next five hours before leaving for the hospital. I spent part of the time on the couch downstairs loving and cuddling our puppy Terri. We arrived at the hospital at 6:00 A.M. the following morning and went down endless corridors to the outpatient surgery department. The woman at the desk asked me what size feet I have. I could not imagine what my foot size had to do with breast cancer, but I told her they were size 7½, and she gave me a pair of hospital slippers. We went down to my room and waited.

Rebecca came in soon after 6:00, wearing a flowing red Indian print dress and looking lovely and radiant. She brought along a woven basket with two of her healing crystals on loan to me for the surgery. She was wonderful, talking with me and holding my hand, and she thanked me for including her in my ordeal since she had felt shut out of Peter and Marcela's pain.

After what seemed like a long time, the nurses finally came in and wheeled me down to the anesthesiologist. The hospital personnel were fussing about not having the new consent form for "Lumpectomy and Lymph Node Dissection" instead of the old plan of "Lump Biopsy." I told them I had signed the new one, but they waited until they heard verification directly from Dr. Orr before they would start the intravenous medication. A nervous woman was beside me, and when the anesthesiologist was finally told to start us both, I told him, "Do her first!" She laughed, and they did me first.

The caring manner of the anesthesiologist helped put me at ease. He said his daughter had undergone this very same operation five years ago, and she was fine. The local anesthetic before inserting the IV really hurt, but I only felt pressure when they put in the IV itself, just as promised. He put a relaxer into the IV, and I began to get drowsy as they wheeled me into the operating room and transferred me onto the table. Dr. Orr was there waiting, cheerful and dressed in his green surgical outfit. I joked with him that while I was under anesthesia anyway, I wanted him to take out the fibroid in my uterus and the warts on my face and to liposuction a couple pounds of fat from my stomach area and put it into my breasts. He laughed and said it would be hard to do all that in the same incision. Within seconds of starting the general anesthesia, I was unconscious.

Because the lump was close to the surface, Dr. Orr elected to take out a small wedge of skin with it and several millimeters of normal tissue in all directions to decrease chances of cutting into any cancerous tissue, which might release cancer cells into my bloodstream. He removed a portion of skin in line with the crease at the base of my left breast 3.0 × 0.8 cm with underlying breast tissue measuring 5 × 4 × 1 cm containing the firm, tan 1.5 cm lump. The frozen section they sent to the lab on the spot confirmed that the lump was a carcinoma.

Dr. Orr was worried because many of my lymph nodes were enlarged and firm, possibly consistent with metastases since the lymph nodes are one of the primary defenses against cancer and usually the first place secondary tumors lodge. He removed the entire chain of lymph nodes in my left armpit in two stages, first the level I and II lymph nodes in a wedge of fatty tissue 9 × 5 × 3.5 cm and then the level III lymph nodes in an additional piece of fatty tissue measuring 3.0 × 3.0 × 0.5 cm Since the lymphatic system returns interstitial fluid to the bloodstream to prevent swelling, and the lymph nodes had all been removed from my armpit, to prevent edema, he installed a drainage tube that went four inches into my arm through a separate stab wound 4 cm below the armpit incision. He closed the two-inch breast incision with sutures and the four-inch armpit incision with staples. I came to consciousness in the recovery room a few hours later with fire and agony in my left armpit, feeling horrible, aching in my entire being. As the nurse wheeled me back to my room, every little bump hurt like hell. Soli and Rebecca were waiting in the room and came to hold my hands, one on each side. My eyes could not focus properly, and I was drowsy, but Rebecca marveled at how alert I was and how good I looked, just the opposite of what I felt. Her positive words encouraged me as I could not do any positive thinking for myself at that moment.

The nurses asked whether I wanted a pill or a shot for the pain, and I chose the injection because it was quicker. Within a few seconds after the Demerol entered my bloodstream, I felt waves of relief, and the terrible pain in my armpit subsided to a dull ache. I floated into a trance-like state of drowsiness from the pain medication every time someone was not talking to me or taking my blood pressure.

Part way through the day, I caught on that I was supposed to urinate in the bathroom "potty seat" with my name on it so the nurses could measure how much liquid was coming out of me. I wondered why my roommate did not use the container with her name on it.

Soli stayed right by my side the whole day, holding my hand or touching my shoulder, attending to details and getting me whatever I needed, like more orange juice and another icepack for my armpit. His healing presence was wonderful. By 6:30 P.M. I could see that he was exhausted and suggested he go home and eat supper. I felt a little afraid to be alone with my aching body, but I knew he needed rest, and I had nurses to help me.

I was delighted when Aida, another therapist on our team, came in to see me that evening, bringing me warm hugs from everyone in the mental health department of the clinic. She brought me a tape recorder Rebecca sent so I could play soft, relaxing music to drown out the noise of the incessant TV of my roommate. My supervisor came by later that night, and I could see she was shaken up by my illness. I was still on an emotional high from the experience of feeling my soul being healed the previous night and was also on a drug high from the Compazine, Percocet, Vistaril, and Demerol the nurses were feeding me, and I babbled on to my supervisor about how having cancer was going to bring marvelous changes in my life and a richness and fullness I never had before. I glowed as I shared with her my joy about the soul healing I had felt in meditation the night before. She talked about how she also worked too hard, and as she left, she said, "If you have any pointers for me, tell me."

In the middle of the night, sorrow hit me when I suddenly realized that my roommate was a quadriplegic. As she was leaving to go home the next day, the cap on her urine bag accidentally came off and soaked everything with urine: pants, socks, shoes, bed, and floor. Getting her dressed had been a two-hour ordeal, so her mother decided to take her home wet and fix her when they arrived. My roommate herself was cheery the whole time, and I did not see how she could cope with a handicap like hers. I felt lucky to only have cancer, something treatable.

Before I left that second day, my new friend Hester, a social worker at the hospital, stopped by to see me. I felt her loving emotional support, but I also felt disconcerted because I thought I saw the fear of my death in her eyes as she talked. She said my osteopath friend Marie was really worried about me, and I assured her I would be fine. She talked about support groups down the line and other things she had for me when I was ready and said that even though she did not know Soli and me well, she really liked us.

I could hardly wait for 5:30 to come for Soli to take me home again; each of the few minutes he was late seemed like an hour. I felt like an invalid because I needed a wheelchair to get to the front entrance and limped into the car.

That night Marie came to our house and cooked me a delicious soup made with shiitake mushrooms to build up my immune system for the chemotherapy.[1] She said that if she were me, she would not eat any dairy products since breast cancers thrive on estrogen, which is naturally high in dairy but also artificially elevated because of the hormones given cows to make them produce more milk. I decided on the spot to quit eating dairy products.

I could see how concerned Marie was, and when I asked why she seemed so worried, she said that just two days before she heard about me, she suddenly got the thought that she herself could get breast cancer. She realized she did not rest enough and feared for her own health, and she knew the insidious nature of the disease of breast cancer from her practice as a doctor. I tried to convince her I would survive, but grief hung heavy in the air.

After dinner, Marie held my right hand and Soli held my left hand, and we all cried together as we listened to the tape of the psychic reading Sara had done the night before surgery and had sent me. I burst into tears as the music at the end of the tape reached directly into my wounded body and soul:

> "You are my hiding place, Oh Lord.
> You save me in my distress.
> You surround my soul with cries of deliverance."

---

[1]Asian medicine has developed sophisticated knowledge of the pharmacological properties of herbs and foods and teaches that eating shiitake mushrooms enhances the function of the immune system.

CHAPTER 8

Staging the Cancer

When I went to sleep that night, the emotional impact of cancer hit my dream world, and I had nightmares about being weak, sick, and tired:

August 17, 1991, Dream:

I am in a Sunday school room, and one of the pupils wants something fired in the kiln. I try to carry the glaze buckets and trays with tiles on them to get ready to fire, but I am so weak and tired, I cannot make it. I cry to them, "I have Breast Cancer." They do not seem to register the fact, so later, I tell people again.

I am trying to rest in a room alone, clad in nightwear. Someone else is in the room, so I go to a different room to be alone; however, then I see I am in a room right outside the doors of the church service, and the churchgoers will be coming out any minute. I go back to the crowded room because I don't want to be seen by the church people in my weakened condition.

Another therapist from work is there. She doesn't know. I tell her, "I have Breast Cancer," and she holds my hand.

I am under the covers with Soli, and I touch his penis. I am aware of his sexuality even though I am too weak to have sex.

Physical agony awoke me at 4:30 A.M., and I cried to Soli, *"My body hurts."* During the night I got a clear internal sense that my doctoral thesis should be shifted to a study of my own healing. I decided to throw out my former concept of doing healing experiments on others and instead to gather all the healing practices I could find to see what results they had on my own body, tracking the effects with careful observation and Kirlian photography. I thought of checking the effect of the anti-depressant drug Prozac for depression, of acupuncture, massage, meditation, special diet, and Mom's doctor that used nutritional supplements. I wanted to try them all! I also felt strongly that Dr. Dianne Skafte from Pacifica should be my core advisor both because breast cancer is a women's issue and because of her compassionate, loving nature.

My period started, and the glow of surviving surgery and the high I got from all the spiritual support that rallied around me wore off: I felt depressed. I cried easily, and my arm really hurt.

The following night I dreamed of a fire, as I was deep into the flames of my own initiation and unsure how much I would be damaged in the process:

August 18, 1991, Dream:

My friend Sara and I look out her window and see smoke and fire. The hill outside has brush fires on it. Quick! Call somebody!

I debate whether to get out of there fast on my motorcycle or stay and fight the fire. I decide to find some kind of implement to put out the fire. I locate flexible snow shovels and get one for me and one for another person to go squash out the fire. I hope that I will not get damaged from smoke and heat inhalation.

Though I felt frightened in the dream, I did not run from the flames of the initiation. Instead, I looked for a way to survive the fire, perhaps a metaphor for finding ways to deal with the side effects of the drastic medical treatments to combat cancer.

The next day was Sunday, and Marie came over again. As we held hands at the supper table, she noticed that my left side had no vitality in it, and she said I needed an osteopathic craniosacral manipulation treatment.[1] Though I had never heard of this healing modality before

---

[1]Craniosacral therapy works with the meningeal membranes, the bones, the connective tissue, and the cerebrospinal fluid to correct blockages which impede the natural flow of healing energy throughout the body (Upledger & Vredevoogd, 1983). For a more complete definition of cranial osteopathy, see Appendix A.

and never understood what the weekend training workshops Marie was always going to were about, I was eager to let her work on me.

First she sat by my left side and held my palm with one hand and my neck with the other. Gradually, feeling crept back into my arm. She held my hands crossed with hers and later held my head in her hands, moving it ever so gently. I visualized angels holding her hands telling them just what to do, and the flow of healing energy into my body was marvelous!

The effect of this simple treatment amazed me as I sat in stillness after she finished, relishing the increase of life energy in my body. My whole left breast felt as if it were having a gigantic breast feeding let-down reflex, with particular intensity in the wounded spot. I felt the healing energy in the breast and the armpit shrinking the pain, and I threw away my constipating Roxicet narcotic painkiller tablets and Tylenol and from then on used only the homeopathic pain remedy, Arnica Montana, that Marie gave me.

Marie said she did not know exactly why she did each intervention she was doing; she just knew to do it from her training. She witnessed the body's processes and actually needed to do very little to get the body's healing energy to flow properly.

From that one treatment Marie did, the numbness in my left arm partially subsided, and its strength increased in my estimate from roughly 15% of normal to around 50% of normal. My arm could hang more normally than before; I had been hunching up because the weight of the arm pulling on my armpit caused pain. Also, my overall body energy level soared. I was delighted with the effects of this simple, gentle, noninvasive, painless healing tool.

After Marie left, I began reading the book she brought me titled *Food and Healing* (Colbin, 1986) and was impressed by the practical holistic concepts in it, teaching how to use food as a healing tool and how to improve my total health by paying more attention to what I ate. I noticed that since I had eliminated dairy products from my diet, my craving for sweets had subsided. Having cancer stimulated me to examine my diet more closely to see whether I could make any changes in the types of food I was eating to improve my overall health.

When Soli went back to work on Monday morning, I meditated and gave thanks for the wonderful things in my life. However, I felt apprehensive about the approach of hurricane Bob, which was on its way to New England. My feelings went up and down, as shown in this entry from my journal:

August 19, 1991, Journal:

I vacillate between being able to hold on to thanking God for my healthy body and getting scared about the cancer disease. At times I fall into a helpless feeling, as if cancer had invaded me of its own accord and might rob me of my life. Then I hang on to my faith and put in the strong thought that my will is in control of my life, and I can absolve this cancer just as I allowed it into me. I can feel God's healing hand within me as I get quiet.

I wept this morning as I meditated and then said aloud the affirmation, "I love myself."

Still, I am scared to die. I do not want to leave this beautiful place and the people who love me so much here, and the ones I love so much. My work here is not yet finished, and my children need me to guide them, Vivien especially.

My heart grieves in its depths at the tragedy of cancer striking me. I ask, "Why, when I am so healthy?" Soli says I already have the answer, to work less. But what about the drive that drove me to work so hard in the first place? I can't just tell it to go away. It has to be dealt with. The soul healing was the beginning, but I see that the road will be long and hard, and I don't know how sick I will be along the way.

I feel this shamanic initiation is a chance to turn around some negative patterns I have been caught in. Everything is in place now, family structure, a loving husband, career advancement, financial security, and a life-and-death crisis necessity to do the deep inner work I need.

Lord Jesus Christ, please help me see the path clearly. I need your Peace and your Presence to stabilize me when it seems the world is falling apart, with the poisoning of Lake Shasta, the toxic train derailment at Seacliff, and just now the imminent hurricane.

By noon, the high winds had me frightened, and possible tornadoes were forecast. I was alone out in the country, and in my weakened state, I knew I would be in trouble if a tornado hit our area. I worried about Soli's safety, trying to drive home in the storm, and felt helpless. I called my neighbors and asked them to watch out for me. Soli's work closed early, and by the time he got home at 1:30 P.M., the winds had escalated and were blowing branches all over. We cuddled up together on the fold-out couch bed in our living room to wait out the storm. The electricity went out, and while we still had phone lines up, I called my school in California and talked to Dr. Skafte, filling her in on the events of the past five days and presenting the idea of my thesis to her. She was wholeheartedly supportive and said she felt honored to be asked to be my advisor. She noted that Soli and I being nestled together with the storm raging outside was an apt metaphor for our togetherness in the midst of the fury of the battle with cancer, which felt like a hurricane tearing through my life. In the evening, during the blackout from the storm, I meditated holding a special multifaceted crystal I had brought from Athens and asked, "What is the meaning of this cancer in my life?" The following visualization came to me:

August 19, 1991, Meditation:

I see three pointed figures, the one in the middle larger. They approach and become three hooded figures who part their hoods, and I see three wise, radiant people with long beards. Their names are Simon, Barabas, and Peter. They come to me and join with our eight hands touching in the center, forming a foursome cross with me at the south pole, the large center figure at the north pole, and the other two at east and west.

They say, "This is about bringing you Wisdom." They enter into my body through the opening in my left armpit where they took out my lymph nodes and go into my center.

"You have Wisdom now and all will see it. Others will listen to your light because you have suffered: people are drawn to suffering." I see light flowing from the center of my belly. The figures hold a collection of beautiful glowing crystals that give off iridescent colored light, green and blue, which coalesces and goes into my third eye, forming a beautiful kaleidoscope effect in my inner field of vision.

I ask, "How will I know how to proceed along the way?" They say that all I need to know will be given as needed. I ask, "Will this end in death?"

They act as if the question were ridiculous and say, "No, it is about Life!"

They say the cancer is a homeopathic cure to bring me to Life, fullness and richness of life. I will be healed, and my job is to witness my healing and document it.

They say with gusto, "We (meaning the four of us) are stronger than cancer!"

In the beginning of the visualization, the cross symbol that the figures formed with me was a mandala image, a universal symbol of wholeness that Jung (1964) found represented in many cultures. The path of healing was leading me to wholeness.

Wisdom entered my body right through my wound, consistent with the shamanic principle of "the wounded healer," learning the secrets of healing firsthand by a direct experience of illness and physical suffering instead of learning about healing by reading books or experimenting on others. A Greek god of healing, Chiron, had himself suffered an incurable wound, and from this personal woundedness, he learned to teach the healing arts to others.

The visualization figures describe cancer as a homeopathic cure, an appropriate comparison since homeopathy introduces into the body's system a tiny amount of the very substance one wants to eliminate so that defenses against this substance will be activated. My body allowed into my system a little bit of cancer, a little bit of death, to activate the forces of Life within me. This meditation brought me emotional comfort, quieting my fears of death, and gave me the concrete job of documenting the process I was going through.

The following morning, I recorded a dream, which was prophetic of the bumpy emotional and physical roller coaster I would be on for the next nine months of cancer treatment.

August 20, 1991, Dream:

I offer to drive a cute little boy, very smart, home to Providence. I check my maps and head out of the cloverleaf hub, taking an offshoot to get onto the main highway. We wind up in a tourist trap, a graveyard type enclosure that has very bumpy roads. We have to go terribly slow. Then steep rocks are at the end. The journey is frustrating, but also like roller coaster and bumper car fun. People are trying to get through but cannot find any outlets.

We stop and look at a map and see only one side outlet leads back onto the highway. Isabel spots it, and we go out the unused gate marked for "visitors." Finally we are headed down the road again.

The city's name is a metaphor, because the word Providence means "**a:** divine guidance or care **b:** God conceived as the power sustaining and guiding human destiny" (Webster, 1965). In the dream, I am taking my inner child home to the divine care of God.

"A graveyard type enclosure" is an apt frame for cancer treatment, because the presence of death is never felt as very far away; however, an attitude focused on death forms a mental trap. The treatments themselves, particularly chemotherapy, feel a bit like death since drugs that kill cells are deliberately introduced into the body, making the physical course of cancer treatment very bumpy.

The emotional course of cancer treatment is also steep, very up and down because of the mood swings induced by the drugs used in the body. The anti-nausea drug used with chemotherapy, Zofran, blocks the neurotransmitter serotonin, an action that is the exact opposite of the new anti-depressant drugs like Prozac, which increase serotonin. So perhaps Zofran might add to the depression that seems to come along with the cancer treatment package.

The dream points to the need to proceed slowly, the opposite of my previous hectic pace, and meditation provided me a way to slow down my mind so my body could heal. Some people get stuck on the cancer roller coaster and never find their way back out. Yet roller coasters and bumper cars are great fun at amusement parks, so the dream says the challenge of the journey also contains an aspect of fun, an attitude that leads to finding an exit from the cancer trap.

The inner part of me that located the way out was Isabel, the person who moved me to check for the lump in the first place. She inspired me to attend to my physical body to check what was going on in it, and this attitude of attention to the physical body helps me get through the graveyard trap and back to God's power sustaining and guiding my destiny, back on my way to Providence. I am a "visitor" to cancer treatment, not a permanent resident.

The morning after this dream, I decided to have a little fun checking the effects of meditation, a little time "home" with Providence, on my wounded arm, testing my arm's strength by pushing on the bathroom scales to see how many pounds of pressure I could register:

11:00 A.M. Test	LEFT	RIGHT
Pushing with arm	28 lb.	62 lb.
Sustaining hand squeeze	40 lb.	60 lb.
Pushing with elbow	15 lb.	35 lb.

Then I meditated at 2:00 P.M. and felt myself being cradled in the lap of the Mother Goddess, in her hands, being loved by her. I felt her presence filling up the meditation room, holding me in her energy, and going on to fill up the whole world, cradling me in its center.

I felt Lord Jesus Christ come and put one hand on my left hand and another on my neck, just like Marie had done, and I felt healing energy flow through my arm.

Afterwards, the results changed to the following:

2:40 P.M. Test	LEFT	RIGHT
Pushing with arm	32 lb.	46 lb.
Sustaining hand squeeze	46 lb.	48 lb.
Pushing with elbow	26 lb.	45 lb.

These results surprised me. I had expected that my left side would get stronger after meditating, but I had not expected my right side to get weaker.

I did the test again when writing this chapter to get a baseline now that my body has healed, with the following data:

Test	LEFT	RIGHT
Pushing with arm	47 lb.	46 lb.
Sustaining hand squeeze	54 lb.	54 lb.
Pushing with elbow	36 lb.	36 lb.

My present strength in squeezing my hands is evenly balanced, at 54 lb. on both sides. However, when my left side was weaker than normal, 40 lb., the right side was stronger than normal, 60 lb. After meditation, the sides got more even, with the wounded arm increasing its squeezing power to 46 lb. and the right side coming down to 48 lb. These figures suggest the possibility that when one energy channel in the body is blocked by physical damage, the body diverts that energy to another place. Meditation may allow the wounded channels to heal, opening them up and rerouting the overflow of energy back into the proper path, balancing and restoring the natural flow of the life force.

The visiting nurse who came to change my bandage that afternoon told me, "I know you are going to be okay. This is very unprofessional to say anything like this, but it's just a feeling I have. I sense which patients are going to make it and which ones I should be concerned about. I know you will be fine. I said to myself, 'There's nothing wrong with this lady.'"

My son Robin called, worried about me. With concern in his voice, he asked if the bag I was wearing under my armpit was for life. He was relieved to hear they would be taking it out at the end of the week.

My dreams and meditations during the next two days brought up the topic of past love relationships that had been lost. As I faced the pain of those losses, I wept and forgave everything, willing to release the men involved to their own paths and asking God to bless them richly. I willed to let go of my emotional attachment to them to seal up the emotional energy that was leaking out of me from those heartaches. I needed all of the strength I could muster to survive the cancer treatments.

I thanked God profusely for bringing Soli into my life, a man who wanted to stay with me, even through the ordeal of cancer, and asked the divine force to bring all of my emotional energy into the present, to my love relationship with Soli and to the budding love relationship with my own body and soul. That night I dreamed I was having a baby, new life within me.

The emotional road continued to be bumpy:

August 21, 1991, Journal:

Depression is here again. Profound sadness is inside me. I fight with the feeling that I might die. Thoughts come to me like, "What if all this positive thinking doesn't work?" and "What if they find that my whole body is full of little cancers?" and "What if something else goes wrong, like getting dysentery or a high fever reaction from the surgery?" and "It's great that I have $90,000 worth of life insurance, in case I die."

Sadness wells up within my soul. I can't explain it or put a finger on it, but it sits on me like an elephant. I am so sorry that my body is sick. My back aches between my shoulder blades, probably from the funny posture of not being able to carry my arm right. My armpit alternately hurts and itches. The muscles in my arms and shoulders hurt, and when I make a new kind of movement, they sometimes stab me with sharp pain. I do not have proper use of my left arm, which feels numb down the back side of the arm, right down to my fingertips. It feels as if it is dying. I am afraid to do anything with my arm for fear I will break something and the J-drain bottle will fill up with blood.

I feel guilty about marrying Soli and then finding cancer. I thought he was getting a healthy, vibrant, dynamic person, and now here is this depressed invalid. I am afraid that I will die, leaving him alone and feeling like he got a bum deal. I want so much to stay here on earth and live a long life, to have 40 years together with Soli at least. I don't want to repeat the Ken Wilber (1991) story. [Wilbur's wife, Treya, was diagnosed with breast cancer on their honeymoon and died from this disease five years later.]

I don't want to be sick! I hate this sickness! I want health and happiness. And I *hate* this depression. Tuesday morning I felt great, rested well, and enjoyed the peacefulness here. Today it seems dreary instead, and fleas from our dog are attacking me, after my blood.

Just now, I don't feel strong and positive; I feel weak and depressed. God, please help me! Have mercy on my soul!

Following this bout of depression, help came from an unexpected source. I had insomnia and was still awake at 3:30 A.M., when I drifted into the following lucid dream:

August 22, 1991, Dream:

I realize that I have no memory of the party the previous night. I tell someone that because I took one sip of alcohol at the beginning of the party, I lost memory of the pleasure of the entire event. I complain about not being able to "get away with" absolutely *anything* anymore. I cannot drink any alcohol, cannot use caffeine, cannot eat refined sugars, cannot overwork my body, and can't even eat dairy products now.

Suddenly three dream figures who call themselves Lightworkers appear. They talk with me and realize my needs. They take me out in the large barn-factory and put me on a table and do bodywork on me. I literally feel their hands on me, working with my body. They massage the third eye point on my forehead with a circular motion for a long time. They also work out the sides of my forehead, particularly on the right temple, where I hear and feel something like a bubble inside pop and release.

One of the men who works on me is young and looks as if he comes from India. He has shoulder length black hair. I feel sexual desire for him, but say nothing. He is my healer.

They also work on my spine. One of them holds me by my head, dangling the rest of my body to release the pressure on my spine. I think I should be sure to smile during the process so others who see me will know I am being helped by the procedure.

At first, I think these people are not human because they have the ability to make the rest of their bodies invisible and only the hands show, so they can freely move around my body to work on me without their torsos getting in the way.

They say they will leave some tea for me to drink, a kind to help me heal. It comes from Brazil and starts with the letters "ch." I ask if the name is chamomile, and the blond one says, "No! You are exhausted, and if you drink chamomile, you will get even more limp-like." They leave several boxes of food, one for me and some for others. Mine has little tins of food to put into my lunches. Most of it is a lentil mixture, which has a few sparse shreds of cheese on top. I wonder about this re-introduction of dairy products into my diet.

Later I sit down with this man, named "Angel." I ask to get to know him a bit. I inquire how old he is, thinking about one hundred. First he says between fifty and one hundred, then well, actually, exactly a century! I joke that he doesn't look a day over fifty, then tell the truth, that he actually looks about twenty-five. He has medium long blond hair and wears a peasant cloak.

He says that at one point he needed an upper left forearm transplant, and his children donated tissue, very painful. He shows me the arm muscle now, and it is very large, strong, and healthy. I ask if he is mortal, and he affirms, "Yes!"

My favorite worker is a slightly older man, Gene, and I feel delighted when they return to do more work on me later.

After being adjusted by the Lightworkers in the dream, I slept peacefully until 9:00 in the morning. Were they angels? The bubble they popped inside my right temple lifted the depression from the previous day, and the physical effects of the treatment they gave me were healing. I felt peaceful inside, not driven to work, but also able to attend to light housework that needed to be done. I looked up the names of teas for cancer in my herbal book and found that the three kinds recommended were taheebo, which comes from Brazil, red clover, and chaparral, all blood purifiers. The next time I went to the health food store, I got some chaparral tea, and for good measure, I got the other two kinds also. I decided to eat a small quantity of a few dairy products so I would not feel so emotionally deprived.

In the afternoon I sunbathed almost two hours, sleeping part of the time and then refreshed, starting to read a book by Dr. Deepak Chopra (1988). I could feel nature's beauty from the lake and the woods healing my soul. I enjoyed the sun and sweat on my body and began to feel like myself again, relaxing into being instead of constant doing.

That night the roller coaster went back down, and I awoke at 2:30 A.M. unable to get back to sleep, worried about the appointment with the surgeon at 8:30 A.M. to find out what stage the cancer had progressed to. To endure the stress of waiting, I documented my feelings in my journal:

What will they tell me about this foreign invader in my body? I still think the lab tests will say that there is no cancer in me, that this is all just a bad mistake. I am too healthy to be sick! I think the lab technicians read my cytology incorrectly.

The opposite thought also passes through, that they will tell me the tumor has spread incurably, and I am an inner mass of cancers eating away at me, on my way out death's door.

I tried to meditate and turn over my thoughts and feelings to Lord Jesus Christ and to Mary, but I could not focus my thoughts. I tried a yoga relaxation exercise, but could not get past trying to relax my toes. The rising anxiety within me is difficult to bear.

At breakfast I told Soli that my stomach was a bundle of nerves, and confiding in him took the edge off the unbearable tension enough so I could meditate for a few minutes before leaving for the appointment.

I prayed and felt Lord Jesus Christ and Mary with me. I asked if I had cancer, and they said I did, but they were with me, protecting me. I felt their arms around me, embracing my soul, and I calmed down enough to leave for the appointment.

The bumpy backroads on the way jarred my armpit, and I was acutely aware of the internal torture caused by the J-drain penetrating my chest cavity. I could hardly wait to get the drain out, but I dreaded hearing the lab results. Soli's emotional balance and talk about superficial matters helped me survive my internal emotional mess on the trip to the clinic.

Marie was waiting at Dr. Orr's office to go in with us. I told the nurse I brought along my Support Group, and she laughed and said that was fine.

Dr. Orr came in accompanied by another young doctor and a friendly young nurse and said, "The news is good. There was no cancer in any of the 31 lymph nodes we removed. That's better than I expected."

He then said some other technical things I did not understand and mentioned "carcinoma in situ" being a rare finding. I asked if that meant that other women do not often have this kind of tumor, and he said no, it meant it was rare inside of my breast. Marie seemed to understand everything he was saying.

I felt relief and a moment of disbelief, like "It's too good to be true!" Then the truth sank in that at least I had really had cancer, but evidently it had not spread far. Marie and Soli and I were all very happy.

Dr. Orr asked if I wanted the J-drain in my side removed, and I clamored, *yes!* It had been aching during the whole time we talked. I had Soli hold my right hand because Dr. Orr said it would hurt, and he was right. He cut the suture and gave a big heave, and deep inside my chest cavity, I felt the ache of the past week begin to move. He stopped for a moment, then gave two more long painful strong tugs, and the drain was out of me. Immediately the nature of the soreness inside changed. I was reeling from the agony of the removal, but instantly I felt relief from being rid of this worm that had invaded my body. The nurse put a bandage over the hole where the tube came out, and within a minute I felt healing set in inside my chest.

I asked Dr. Orr how he felt about alternative healing methods like massage and acupuncture. He was in favor of anything I thought might help, adding that the worst it could do would be to get expensive.

He interjected, "You already meditate, right? That is really important. I notice that attitude and meditation are two factors that distinguish people who do well versus those who don't, but I can't say *why* it is so. I can't say why cancer struck in a life like yours that seemed already healthy." I marveled that this traditional physician recognized the healing power of positive thinking and centering the mind in meditation. He knew about Deepak Chopra's Maharishi Ayurveda Health Center for Stress Management and Behavioral Medicine nearby in Lancaster, Massachusetts, and felt their diet and herb recommendations would probably be good for me.

After Dr. Orr left, the nurse gently took out the stitches in my breast. They came out with just a few twinges, but taking out the armpit staples hurt quite a bit. She asked me to raise my arm to get to the ones deep inside, and I could barely tolerate having my arm lifted because of its muscle damage. Finally she got out the last two staples, and my body was free!

On the way home from the doctor's office, Soli and I had a difficult discussion about financial matters. Soli had budget shock from having laid out $10,000 already that month, including college tuition for both of my children as well as my doctoral program, and he feared that I would want to spend too much money on alternative healing methods. He stated, "There's no more your money and my money; it's all one," and asked to combine our checking accounts.

I declared I absolutely needed my own account because my Sagittarian spirit needed freedom and explained, "If you put me into a box, I will whither and die." After we talked about the practicalities of how confusing a joint account would be, Soli agreed to keep separate accounts and found a way to avoid the service charge on my account.

When we arrived home, I took a good look at my scars in the mirror. The two-inch breast incision was nicely hidden in the folds of my breast; in fact, getting a good look at it was difficult. But I was aghast at the appearance of my armpit. The incision was four inches long and had contained 14 staples. The skin was bunched up where it had been held together by the staples and looked like two lips, running side by side about 3/8-inch wide each, with a slight reddish color.

Hair was growing around the lips but not right on them. I thought it looked like a vagina and asked Soli if he would get confused and try to screw my armpit. He exclaimed, "Not a chance!"

My younger sister called soon after we got home to find out the results of the pathology report. We shared our joy at the good news, and I informed her, "Keep me on your Christmas list. I'm planning to stay alive for quite some time yet."

Soli took me sailing on our lake that afternoon, and we had a long, intimate talk. I enjoyed the slow pace of the boat since the wind was low. After sailing, I was very tired and collapsed on the pier, enthralled by watching cloud formations in the sky. They looked like giant Rorschach inkblot tests to me. I pondered how rarely in my life I had held still long enough to really watch clouds. One looked like an embryo, but then I saw it was a woman, with her hair streaming behind her, mouth open in protest, and her breasts flaring open at the ends. I felt for her agony and her plight.

I walked down to get the mail, hoping to find my paycheck, and instead found three get-well cards. I broke into tears at the thoughtfulness of the people who had sent them and was touched by the love in the positive messages they carried.

I listened to the reading from Sara again, and it seemed deeper than before. Again, I cried with the beautiful music at the end, letting the words about Jesus being my hiding place soak into my spirit, relaxing my body and nourishing my soul. I watched the beautiful trees outside waving their leaves in the wind and let nature and the music heal my soul.

The following day I felt much better. I could feel the trees and the lake healing me. When I took off the bandage that had been covering the hole where the J-drain came out, I was alarmed to see that although the hole was only about a quarter of an inch long, it was surrounded by a black and blue spot the size of a half dollar.

I had forgotten to ask the doctor whether or not I could go swimming, so I decided to get into the lake for a short time. To my joy, I could actually do a modified version of the breast stroke. I loved being in the water with the fish, then sunning myself on the dock like a turtle. Time moved more slowly as I enjoyed the path along my life instead of barreling down it full-speed-ahead.

Soli called at 6:00 that evening. He had been helping Marie move to her new apartment, and he invited me to join them for supper at a Chinese restaurant near her new home. I hesitated for one second at the thought of going out into the world again on my own, but then decided I had enough energy to do it. I got dressed up and drove there, excited to feel mobile again.

As I walked into the restaurant, a woman rushed over to me and joyfully threw her arms around me—the visiting nurse who had come to my house earlier in the week. When I told her the tumor was only 1.5 cm and my lymph nodes all came back negative, putting me into Stage I cancer, the most treatable level, she cheered, "See, I *knew* you would be just fine!" Soli came over to see what all the fuss was about. The nurse recognized him from our wedding picture and introduced herself to him, calling him "Mr. Stone," not knowing that Stone is my maiden name. Soli bragged to her about how I went swimming that day and drove myself there. I felt like a miracle!

CHAPTER 9

To Chemo or Not to Chemo?

With no lymph nodes showing any signs of cancer, deciding whether or not to include chemotherapy in my treatment was difficult. Dr. Orr was strongly in favor of giving me chemotherapy, but Soli was opposed. I realized the decision and its consequences were ultimately mine, so I set out to get all the information I could find to make the best decision possible.

The statistics Dr. Orr quoted me were that with radiation treatment only, 20% of the women in my situation would have recurrence of cancer; however, with chemotherapy added, only 10% of the women would have recurrence. He stated, "It's like buying an insurance policy, and worth it in my estimation," adding that the side effects of hair loss and feeling sick were very temporary, and only one out of every 1,500 chemotherapy patients developed leukemia as a side effect. He felt that if the treatment damaged the immune system more than it helped to eliminate the cancer, the statistics would show a decrease in the survival rate for women who chose to get chemo.

Dr. Orr added that all treatments carry risks. "Surgery is a crude method. Chemotherapy is also crude. It's like bombing the whole city just to kill one terrorist." He could give me no 100% guarantees that even with chemotherapy, I would definitely be cancer free. "After any treatment, you have to go on at the end 'as if' you are cured, but there is always a slight chance that cancer might return. Even with the radiation therapy, 20 or 30 years down the road, cancer might recur in the breast or in the bones or brain." He also warned me that the oncologist would quote me statistics that were more pessimistic than the ones he had given me. I realized that medicine could not fully protect me from my fear of death.

Soli did not like the idea of my undergoing chemotherapy, fearing the long-term side effects of the chemicals used, especially since most women with Stage I cancer would not have a recurrence of cancer even without chemo. Soli did not think any more cancer cells were left in my body, so chemotherapy was unnecessary from his viewpoint. He admired my strength and thought I could handle any tiny cancer cells with alternative methods.

As I wrestled with my fear of recurrence of the cancer and my own fears of chemotherapy and the opposing viewpoints of my doctor and my husband, a dream brought a vivid image of chemotherapy:

August 24, 1991, Dream:

———BANG! within my system, blasting out the cancer.

A dead tree trunk with all branches cut off of it extends over the piano, and the woman is afraid to remove it.

Grandpa Bontrager says do it. He brings a chain saw and cuts off a knob on the left side of the piano and pries the brass cover off a circular horn decoration on the front of the piano.

I feel he is ruining the piano, but he says no, what he is doing is making way for a new, better piano.

The dream begins with an image of cancer being present in my body, but being blasted out by a strong treatment, an apt metaphor for the way chemotherapy works to wipe out cancer cells.

The dream image of having a dead tree blocking my piano was a good description of the way I felt about having cancer: the illness seemed to block my access to my piano, my symbolic source of music and creative expression. In the dream, the woman was afraid to remove the dead tree trunk, the piece of death blocking access to a creative life; however, my grandfather, who had been a healer, said to do it. So the dream seems to recommend chemotherapy to remove the cancer from my body.

My main fear of chemotherapy was losing my hair, and the curly hair on my head could be compared with the circular horn decoration on the front of the piano. Losing my hair would be a bit like sawing off the covering on the curly horn on the front of the piano; however, instead of ruining the music in my life, my wise inner healer said that this "chain saw therapy" (chemotherapy) would be making way for a new and better source of music and creative expression within me.

Marie called and felt I should not have chemotherapy because once an unnatural treatment like chemotherapy were introduced into my body, the natural flow of the body's health would be unalterably disturbed, adding, "Once you've let the horse out of the barn, you can't get it back in." She had witnessed this process in patients in her medical practice.

A respected friend who had lost his mother to pancreatic cancer several months earlier was also opposed to my getting chemotherapy. He reasoned that the object was to do everything possible to build up the immune system, not purposely destroy it, since chemo makes the white blood cell count temporarily go down after a treatment.

On the other hand, my parents were in favor of my getting chemotherapy, wanting me to use every opportunity I had to improve my chances for survival, agreeing with Dr. Orr and my dream. My psychiatrist supervisor at work told me emphatically, "You can do all the meditation you want, *but do the chemotherapy too!*"

On August 26, just 11 days after surgery, Soli took me to Deepak Chopra's Maharishi Ayurveda Health Center for a workup by Ayurvedic physicians.[1] The doctor there encouraged me to go through chemotherapy, explaining, "Even though the chemotherapy only improves your survival rate by 10%, if you are one of those 10%, then it is 100% for you."

The physicians there diagnosed my type by examining me and feeling my pulse at many different points on my wrist. My typology turned out to be a combination of the Pitta and Kapha types with Pitta predominant.[2] They gave me diet instructions to balance this particular body type. The Ayurveda Center doctors said that the most important key to health was daily meditation and encouraged me to learn Transcendental Meditation (TM), the form of meditation they

---

[1]Ayurveda is the Indian science of healing and rejuvenation that uses natural substances to enhance health and well-being.

[2]Ayurvedic medicine uses three categories to define the physical and emotional makeup of the body:

VATA: slender body, high metabolic rate, quick and changeable mind, vivacious manner (symbolized by the sparrow)

PITTA: in-between size body, fiery and intense nature (symbolized by the lion)

KAPHA: strong and solid body, serene and loving, prone to weight gain (symbolized by the elephant).

For more on this subject, see *Perfect Health* (Chopra, 1991).

taught and practiced. They also recommended herbs and a daily routine that included drinking hot water with a bit of lemon first thing in the morning to cleanse my system, sesame oil massage, and yoga. My energy suddenly gave out during the last half hour of my appointment there, and I could barely sit in my chair while the attendant explained the yoga postures and other complicated parts of the routine.

I wanted to do everything I could to improve my physical health, so I decided to try everything on their long list of suggestions for me. One of the items they recommended was a monthly purification of the intestines called "panchakarma," which involved taking some ghee (clarified butter) for several mornings in a row and then taking a laxative to flush out the intestinal tract to remove accumulated toxins.

Soli did not want me to do the purging. He felt it would rob me of too many of the nutrients that my body needed to heal from the surgery. He reasoned that purification would be more appropriate for people with diets that were less pure than mine. I did not listen to his advice, however, and tried the monthly cleansing in early September. I had some trouble swallowing the ghee, but it did its job of softening up and flushing out the intestines so well that by the time I took the laxative, my intestines were so empty that nothing was left to flush!

Had I known that I would have constant diarrhea with the radiation treatments, which started a couple of weeks later, I would have skipped this intervention to clean out the colon. I decided not to include panchakarma in my monthly routine because I did not like doing it and also because my body just naturally got diarrhea once in awhile without helping the process along. Once was enough for panchakarma, the only part of the Ayurvedic routine that I did not like.

I found many of the other Ayurvedic suggestions quite helpful. The Ayurvedic consultation changed not only the type of food I was eating, but also the *way* I had been eating it. Before, I had often shoved down a meal on the run, not giving my digestive system a chance to process the food energy in the meal and thus building up residual impurities in my cells called "ama" from incomplete digestion (Chopra, 1991). I learned to eat more slowly and to enjoy my food more.

I began a daily practice of yoga and meditation, the vital keys to building health. I had never done yoga before, and I had been too exhausted at the end of my appointment to get the whole story on how to do the yoga stretches; however, undaunted, I learned the yoga positions from the sheet of postures the Ayurveda Center sent home with me. I felt very proud of myself that I got the yoga routine down so I could get through it in seven minutes flat, very efficiently, and get on to my meditation.

But several months later, when I took the Stress Reduction and Relaxation Class at the University of Massachusetts Medical Center,[3] I was surprised to learn that the purpose of yoga was not to rush through it, but to do it slowly and "mindfully," in order to stretch the muscles and feel the body's energy during and after each posture. *I had been doing yoga just like I used to eat!* I began doing the yoga slowly and with awareness, luxuriating in the wonderful feeling of my body's energy circulating freely from my brain through my spine, out my nerves, and into each organ and muscle of my system.

To get more information on breast cancer, I went to the library and checked out all the books I could find on the topic. I remembered Dr. Orr quoting something about "carcinoma in situ" in my breast, so I read everything I could find on that condition. I discovered that "carcinoma in situ" meant the cancer was still "in place" with a membrane around it, so the tumor could not have spread. The real danger of breast cancer is that it eventually outgrows the milk duct or milk lobule where it starts and invades the surrounding tissue, shedding cancer cells into the lymphatic system, which then circulates the malignant cells throughout the body, starting up tumors called *metastases* in distant locations. None of the books I checked out recommended chemotherapy for "carcinoma in situ," so I rejoiced in the false belief that I would not have to undergo chemo afterall.

The statistics I read shocked me when I realized how low the survival rates dropped with positive lymph nodes:

Survival Rates Based on Lymph Node Status

No Nodes Involved:
Recurrence:
in 18 months 5%
in 5 years 18%
in 10 years 24%

---

[3]This program, developed by Dr. Jon Kabat-Zinn, was featured in one of the segments of the Bill Moyer's PBS special "Healing and the Mind" (February, 1993).

Survival:
5 years 78%
10 years 65%

With 1 to 3 positive nodes:
10-year *survival* is 38%

With 4 or more nodes positive:
10-year *survival* is 13%

Average of all patients:
10-year *survival* rate is 46%

(Spletter, 1982, p. 120)

Cancer is staged according to tumor size, lymph node involvement, and whether or not metastases have been found. My cancer fell into Stage I, the most treatable level, because no nodes were involved, and the tumor measured less than 2 cm in size. Any lump over 2 cm automatically qualifies as Stage II cancer, whether nodes are involved or not, and any node involvement disqualifies from Stage I. The difficulty of treatment increases as the stages get higher, all the way up to the highest level of Stage IV, where a metastasis is found in another part of the body.

Stage I cancer that is treated only with local treatment has an approximate five-year survival rate of 80%. In Stage II, the five-year survival rate drops to 65%; in Stage III, 40% will survive five years; and only 10% of the women with Stage IV breast cancer will be alive five years later (Love, 1990, p. 235).

The dream I had the night before my visit with the oncologist shows my relief that I had discovered the lump while it was still small enough to fit into Stage I and my faith that I would survive the illness:

August 28, 1991, Dream:

In a group a woman says she is going to have a lump removed from her breast soon. I move forward to tell her that she is in a lower risk group because she is so young. I tell her that my lump was cancer, but everything is fine because they got it in time.

In the waking world, the prognosis for a younger woman is generally *worse* than for an older woman, since the cancer found in a younger, more vigorous body is likely to have more energy and thus be more aggressive. However, in the dream world, youth is considered an asset to survival, perhaps because the youthful attitudes of looking at the world with freshness, enthusiasm, and vigor might contribute to lowering the risk of recurrence. In the dream I am confident that I will survive breast cancer because I had caught the illness in its early stage.

I arrived for my appointment with the oncologist confident that I would not need chemotherapy and armed with a list of twenty questions for him. He advised me to have six months of chemotherapy, and I asked him what theory was behind his recommendation since the cancer was "in situ" and therefore could not have spread. To my chagrin, he informed me that my "in situ" cancers were found around the main tumor, which had already broken out of the milk duct and had invaded the surrounding tissue.

My heart fell, and with shock, I read the pathology report right before my eyes:

> Poorly differentiated invasive ductal carcinoma with atypical ductal hyperplasia and rare duct with intraductal carcinoma present outside the main tumor mass.

Dr. Orr had told me exactly what the tumor was, but my mind had been clouded by my high level of anxiety at the time, and the only thing I had remembered from the "technical things" he mentioned was "carcinoma in situ." Both Soli and Marie remembered him saying that the main tumor was invasive.

I had to look up the meanings of the rest of those strange, technical words. *Poorly differentiated* cells look wild and are usually more aggressive than well differentiated cells, which are closer to normal cells (Love, 1990), so my more primitive cells carried negative prognostic weight. *Ductal* means that the cancer started in the milk ducts in the breast instead of the milk-producing lobules. *Hyperplasia* is too many cells in the duct, and *atypia* means that some of those extra cells look strange, both precancerous conditions found in parts of the biopsy tissue around my main tumor. The *rare intraductal carcinomas* found were separate cancers forming, which had not yet broken out of the milk ducts, "carcinomas in situ." About 30% of the women who die of other causes and are autopsied have precancerous hyperplasia or atypical hyperplasia, so probably many women are "walking around with these conditions, and we don't know it because we're fine, we have no reason to have biopsies, and they don't show on mammograms" (Love, 1990, p. 192).

The surgeon had expected to find metastases since the lymph nodes taken out from my armpit ranged in size from 0.3 cm to 1.5 cm, with several of them being as large as the tumor. These nodes might have been enlarged because they were working overtime to gobble up all the cancer cells that the lymphatic system was shedding into them from my breast.

I asked the oncologist, "If there were tiny metastases, couldn't a strong immune system handle them?" He laughed at me and inquired how the cancer got started in the first place if the immune system were strong. Standing firmly on the side of caution, he added that even though my armpit nodes were negative, some cancer could have gone to the nodes in the middle part of the body, where they could not be taken out, or could have escaped the lymph nodes and gone directly to bones or organs.

I asked how I would know if a tumor were in some other part of my body, and he reported that I would not know until the cancer began to cause damage. The statistics the oncologist quoted me were that the incidence of recurrence in my group was 30% and that chemotherapy helped only 25% of those 30%. Even though he thought chemotherapy might start me into menopause and said I was likely to lose my hair, he felt that the survival odds for any tumor over 1 cm were better with chemotherapy.[4] Since the tumor was invasive, a

---

[4]Another reason that I needed chemotherapy was that the breast tumor was estrogen-receptor-negative; in other words, the cancer cells in the breast tumor did not have receptor sites for the hormone estrogen. Tumors that *are* sensitive to estrogen or progesterone usually grow more slowly, so the overall prognosis for those tumors is better. In addition, if the estrogen-receptor test is positive, the tumor can be treated with the hormonal drug tamoxifen, which blocks the estrogen going into the breast and therefore would tend to stop the growth of malignant tumors sensitive to estrogen. My tumor fit the general rule that premenopausal women usually have the more aggressive estrogen-receptor-negative tumors, and postmenopausal women are more likely to have tumors that are estrogen-receptor-positive (Love, 1990). In general, postmenopausal women who have breast cancer are usually not helped by the regular chemotherapy treatments, but premenopausal women have a higher survival rate if they do get chemotherapy. Tamoxifen is technically a form of chemotherapy too since it is a chemical which is a therapy, chemo-therapy; however, it has far fewer side effects than the harsher forms of chemotherapy offered to most cancer patients.

systemic treatment was important in his professional opinion: "You're young, and you should have every chance you can get."

To see whether I might already have metastases, he sent me downstairs afterwards to have a chest X-ray to look for tumors in my lungs, and bloodwork to check for damage to my liver and other organs. My stomach was churning from shock and dread of six months of needles and drugs in my veins. After the X-ray, I nearly fainted when the lab technician drew my blood to check my liver profile. I sobbed to her, "I didn't think I would need chemotherapy, and now it looks like I have to have it."

With real compassion in her voice, she declared, "Oh, you poor thing," and called Soli down from eating his lunch.

My body felt very weak. I thought, "First they take things out of my body, then they put other things in." The concrete thought of chemotherapy absolutely knocked my feelings flat, but I wanted to give myself every chance I could to live. When Soli arrived, I held his hand and cried on his comforting shoulder.

Soli said, "It looks like it's going to be harder than we thought."

Since my mind was full of thoughts of metastases, chemicals, throwing up, hair falling out, and death, getting back into the mode of positive thinking was difficult. If I skipped the chemotherapy and got a recurrence later and were dying before the end of my natural life span because the cancer had spread before it could be detected, I would have trouble forgiving myself for not getting chemo.

Part of me wanted to die right then so I would not have to face all of the physical pain coming up, but that part was overshadowed by my love for my children and for Soli. I knew my work here was not done yet.

When I got home from the appointment with the oncologist, in my mailbox was the gift of another reading from my friend Sara, done before I had talked with the oncologist, which advised me to get the chemotherapy to kill any leftover cancer cells and to relieve my worry about getting a recurrence. She felt that I could add alternative healing measures to counteract the negative side effects of chemotherapy and to regenerate after the treatments were through. Once again, she talked about how this illness was going to bring joy and connection into my life, a new relationship with others, and a new look to my own life, a new way of being in the world.

Uncontrollable sobs broke out after hearing the reading, because I realized it was right: if I did not do the chemotherapy, I would be sentencing myself to living in fear and doubt, attitudes that depress immune system functioning and would encourage a recurrence. Soli had gone back to work, and I was home alone. A dear friend was on her way over to visit me, and I could hardly wait for her arrival. When she knocked on my door, I greeted her with a flood of tears and jumbled up words: invasive tumor—chemotherapy—hair falling out—etc. I was very grateful for her healing presence with me.

The strongest emotional conflict I had was losing my hair, which I had never really ever been satisfied with in the first place. It was too thin and straight. As I meditated and worked through feelings of loss about getting a treatment that would make my hair fall out, I felt that the topic of my thesis should be something like *Cancer as Initiation* to honor the process that was going through me.

My heart was touched by a beautiful bouquet of flowers that arrived from the agency where I worked. Rebecca called to check how I was doing and said I should "savor the slowness" of this healing time. She also said that Peter's widow Marcela, who was a psychotherapist just like Rebecca and me, was suicidal and had become very angry with Rebecca for arranging a psychiatric hospitalization for her. Marcela said to Rebecca, "Don't you call my doctor one more time, and don't you try to get me into a hospital again!" Family and friends were keeping a 24-hour watch on Marcela, who had signed up for electroshock therapy.

My older sister sent me a whole case of aloe vera juice and called me on the phone. She told me, "I know you are healed already. Your soul is healed, so your body will heal too. I'm holding you in the Light, visualizing your body strong and healthy." I asked her to stick naturally curly hair on me into her visualization, and she laughed.

I stopped by the wig salon and was moved by the human warmth of the owner, whose wife had been through breast cancer eight years earlier. He said that unless they had changed the drugs, I would probably lose 95% of my hair. He encouraged me to get a wig before I started the chemotherapy, warning that losing the hair is very traumatic, and gave me an application from the American Cancer Society to subsidize $125 of the cost.

As I was browsing through the shop looking at all the wigs and getting excited about getting one, I overheard him saying to the others in the salon, "We just go from day to day worrying about how to get enough money to make a living, and then some people have much harder things to face in life."

September 1, 1991, Journal:

Thoughts from meditation today:
I think of going through chemotherapy, and I feel
scared
frightened and
afraid.

My body rocks back and forth, trying to comfort the pain within. Agony is in my soul at the thought of facing the ordeal of putting poison into my veins.

I think of Lord Jesus Christ in the Garden of Gethsemane, knowing full well that on the following day he would die a horrible, painful death on the cross. He had full choice of whether or not to go through with his death and asked God if there were any way to spare him the torment of the cross. However, he also consecrated himself to God's will for his life.

I feel the agony of Jesus, dreading the pain and being afraid of the unknown, feeling death close at hand.

Then, like Jesus, I release my fears and totally consecrate my life to serving my Divine Purpose. As I let go of the struggle, a profound stillness comes into my soul, and my body stops rocking and is filled with peace.

The second day after I had returned to work, the psychiatric nurse pulled me out of session to tell me that Marcela had killed herself. I sent my client home and sat with Rebecca as she sobbed violently. She had stayed with Marcela and Marcela's mother the previous night, and when she left to come to work that morning, she did not know that Marcela had committed suicide there during the night, using a book of instructions that was meant for terminally ill cancer patients, just like her husband Peter had done two months earlier.

Marcela had fooled people by signing up for electroshock therapy the following day and being warm, jovial, and lighthearted the night of her suicide, just like Peter had also done. Rebecca wept, "Her poor, poor Mother!" She also added, "At least I didn't have to find the body this time."

I felt angry with Marcela and said, "She wouldn't let you help her. You tried, and everyone got mad at you for it; she did not want help!"

I felt pissed because I was fighting for my life with all of my energy, and Marcela had a healthy body and destroyed it. Rebecca agreed, "Yes, she had that beautiful, slim body, and she threw it away."

I thought about how the difference between Marcela and me was that my *desire* was to live and hers was to die. The condition of the body might be irrelevant in the long run; the desire is what gets worked at till it is eventually manifested.

CHAPTER 10

Beginning Radiation Therapy

The oncologist wanted me to have the radiation therapy before I did the chemotherapy, and I was happy to comply since I dreaded the chemo and was glad to hang onto my hair a little bit longer.

Undergoing radiation therapy was a choice I made because Dr. Orr gave me the option of going back after the lumpectomy surgery and having a mastectomy instead of doing radiation; however, since the mastectomy cure statistics were only slightly higher than the 80% cure rate on the lumpectomy-radiation combination, I decided to keep my breast. I had just returned from my honeymoon several months earlier and was not willing to part with my breast, even though it had little baby cancers in it.

In my next checkup with Dr. Orr, I inquired why my lump had not shown up on the mammogram two weeks before I found it. He responded that the breast tissue of younger women is so dense that it hides a third to a half of the lumps and that in older women, the breast tissue is replaced with fat, which is easier to see through and hides only about 10% of the lumps. Another possibility was that the lump was located too low in my breast to get included in the picture.

Dr. Orr informed me that radiation does not usually substantially disfigure a small breast like mine; older women with large breasts have more disfiguration, but if they opt for a mastectomy instead, they come out lopsided. Nevertheless, I had a nightmare about having radiation treatment and my nipple winding up on my shoulder where it showed through my sleeveless shirts. The radiation doctor assured me the distortion in my breast would not be that severe!

I asked Dr. Orr how big the "in-situ" tumors around the invasive tumor were, and he reported they were microscopic in size. Medical diagnostic methods had no way of telling whether or not these tumors would turn into aggressive cancers, but the radiation therapy stopped them from growing.

While I was asking Dr. Orr so many questions, I also inquired, "Was it hard for you to tell me that my lump was cancer?"

He responded, "It's never easy to tell that to a patient, but it was not too bad in your case because I knew what you had could be treated. The ones that are really difficult for me are the ones where we have to tell the patient there's really nothing much we can do for them." I thanked him for communicating that message of hope to me at the crucial point of diagnosis, easing the terror I felt from knowing I had cancer.

My emotional life was going through a subtle but profound change. Before I found the cancer, my dreams were often plagued by images of sexual conflict; the following dream paved the way for greater harmony between my internal masculine and feminine aspects:

September 6, 1991, Dream:

A woman is in a strange part of town, looking at restaurants. One sign says, "If you eat fish on a diet, you cook it one way, but if you eat fish for pleasure, you cook it another." Why is she there, she wonders. She decides to go back to familiar territory. As she walks, she notices a man in a car who intends to do her evil. She wants to get out of there FAST and hangs onto the door handle of a car going by, with her body and coat streaming behind.

He follows her, so she decides to levitate and get up so high he cannot reach, and she focuses on that. Somehow it does not completely work, and she winds up in face to face contact with him. Four people are together: she is fight-wrestling with a female, and a male is there standing by. I stand by to protect her if she needs me or if the fight gets unequal. I think about getting into the fight myself, but then decide I don't want to physically fight with anybody there. The man still wants to fight, but asks her now to do it "just for fun."

I say I have another idea, something I just learned. I have us all get into a circle and practice sending each other unconditional love. I say it's great for settling things. We start with this woman and the man directly across from her in the circle, which has about 12 people in it. Then we go in pairs. I tell them, "Just open up your heart and let the love flow out. You will feel a little tingle."

The dream opens up with an image of the strangeness of the new world that was opening up to me, and the feminine within is trying to decide how to cook her fish, debating how to nourish herself. The sign in the dream notes that if one prepares the food for a diet, thinking of the mechanical result the food will have in the body, food is cooked differently than if one eats for the pleasure of eating.

I took care of my body before I got cancer, exercising regularly and not eating red meat or drinking alcohol, not smoking, generally avoiding harmful foods, and tossing vitamins down my throat to ensure adequate nutrition. However, I treated my body like a machine when caring for it. I exercised to force my body into the shape I wanted, not for the pleasure of moving my body. The contrast in the two attitudes is similar to the contrast between doing aerobics to whip the body into shape and doing yoga slowly, with consciousness of the contraction of each muscle and pleasure in the release.

I had been choosing my food for the mechanical result it would have in my body, not for the pleasure of eating, and I usually hurried through my meals, pausing just long enough to stuff some gas into my tank before I sped off to the next activity. Eating for pleasure would imply eating food slowly and savoring each bite, a practice that aids digestion and thus promotes health in the body, decreasing the buildup of the toxic impurities that result from incomplete digestion.

The woman in the dream gets uncomfortable with the new attitude of eating for pleasure instead of to diet (a word whose first three letters are "die") and tries to return to her former mechanistic attitude toward nourishment, familiar territory; however, the male principle within her will not tolerate this return to unconsciousness and goes after her. Initially she perceives him as evil and retreats to her old defenses, first of going so fast that he cannot catch her, and then of levitating, getting so intellectual and spiritual that she will be out of his reach. But neither of these old defense mechanisms works, and she finds herself faced with the internal conflict.

Instead of one side trying to repress or eliminate the other, the dream ego suggests that the opposing attitudes join together and send love to each other, opening up to each other and letting love flow between them.

The meditation following this dream shows a moment of pure joy that resulted from the shift this union between inner factions made in my emotional state:

> *September 6, 1991, Meditation:*
>
> In my mind I hear Lord Jesus Christ say to me,
> "Today you are healed."
> I visualize a mental image of a double helix spiral.
> "I am in your genes.
> Your body is made of Love."
>
> I weep tears of pure joy and marvel at the inner change in my feelings. My soul has been healed, and my heart has been unblocked and opened up to love. True and deep happiness is within me now, where before at my core was sadness. I feel an inner radiance shining out of my face, a glow in my total being. I feel absolutely wonderful!
>
> True, the world is still full of sorrow, but inside of my heart is love for myself, love for my life, and love for others.
>
> I feel love flowing in and out of my heart and feel connectedness to the earth and to the heavens.
>
> I love my life!

Right after surgery, in a moment of feeling fear and some despair about my situation, I had written to Arny Mindell, one of my former analysts, asking for help to understand this strange event in my life. I laughed at the letter he sent me in response:

> Dear One,
>
> Congratulations on your marriage and on the production of a mass of good intentions and messages, which are trying to come out through your wonderful dream body.

He was absolutely right. The cancer was a message. In meditation, I realized that confronting my mortality and letting cancer teach me how to live better had required great courage. In my mind the following thought came to me as if from Jesus:

> *Death is my Brother.*
> *I made friends with him through the cross experience.*
> *He comes with a smiling face to those who have lived well.*

On Friday, the 13th of September, I went to get the radiation markings on my chest. While I was putting on my hospital gown, I saw an elderly gentleman take off his gown in front of me, revealing large marks on his chest a quarter of an inch thick in a one-foot square, like telescope marks. Then the technologist put green magic marker lines all around my left breast to figure out the radiation angles for the machine and took a polaroid picture of the marks. When she then said she was going to tattoo these points permanently into my chest, my heart jumped into my throat, and I panicked, looking for the nearest exit to escape. I did not want to look like the man I had seen when I came in. However, when she assured me the seven marks would only be the size of a pinhead each, I regained my composure.

I thought about writing home to my parents that on Friday the 13th I got tattooed and had my first nude pictures taken!

The technology required for this healing treatment frightened me, and I felt apprehensive about exposing myself to the radiation itself, unsure whether the treatments would cure me or harm me. That night I dreamed about snakes again:

September 14, 1991, Dream:

While I am waiting to see a client, snakes are in the house. I squash little ones. My former husband goes after a big copperhead (banded body) with a three-pronged tail. It escapes.

The enlightened attitude of seeing death as a friend had not yet worked its way into my unconscious, and in the dream, I am squashing snakes, trying to avoid all possibility of death or expanded consciousness. My former husband, the former attitudes of my inner male, was trying to kill the large poisonous snake, trying to totally repress this energy of transformation in an attempt to stay in the old snakeskin, hanging onto my old level of consciousness. Fortunately, the large poisonous copperhead eludes my inner male, leaving my ego in emotional insecurity but also permitting the snake freedom to continue his role as bringer of consciousness.

The three prongs on the snake's tail made the overall shape of the snake into that of a trident, a symbol representing a triple phallus, reinforcing the phallic nature of the serpent symbol. The underground gods Hades and Poseidon often carried a trident as a symbol of their phallic power. In Christianity, their images merged into that of the Christian devil, who was often pictured in the Renaissance period with a three pronged penis (Walker, 1983). Present day Halloween costumes of the devil may show him carrying a trident. This snake was trying to plant new life in me; but inside, the dream shows I am resisting these changes like crazy, trying to squash the snakes.

As I was wrestling with my fears once again in meditation, in my mind I heard the voice of my spiritual mentor saying, "Whether you live or die doesn't matter. The important thing is that you love God." I realized she was right: the correct question was not, "Will I live or will I die?" but "How can I love the God within me?"

The next week I flew out to California again to begin the final year of coursework towards my doctoral program. I felt radiant, eager to tell all my friends of the Master Lesson on "How to Love Yourself, Barbara" that I was receiving and of the inner healing of my soul.

I discovered that when I wore the colors of purple, teal, and turquoise, I felt better, as if the colors themselves were channeling healing energy into my body.[1] So I wore my new clothes I had just bought in these three colors out to California, and I felt beautiful, radiating inner peace and deep joy. Perhaps the reason I felt good wearing the blue-greens and purples is that I needed healing in the areas of my body that corresponded to those colors: green for the chest-breast area to heal, turquoise for the throat to speak out better, and purple for my mind and spirit.

---

[1]Colors activate subtle differences of energy in different locations of the human body, with the warm colors tending to stimulate the lower half of the body and the cool colors the upper half. Black tends to be good for the soles of the feet, and then the color spectrum goes up in order from red to violet from the base of the spine up to the top of the head. Thus red is good for the bottom of the spine, orange for the sexual organs, yellow for the navel area in the solar plexus region, green for the heart area, turquoise or blue for the throat area, indigo or violet for the forehead area (third eye), deep purple for the top of the head, and pure white for the area above the crown of the head (the spiritual center). Pink (a combination of red and white) is also good for the heart area. Stones of a specific color can be worn to help a particular area (Silbey, 1986).

The initial reaction of most of my friends who had not heard I had breast cancer was a sad darkening of the eyes, clouding over with sorrow and perhaps fear of my death. But the change in me was a cause for celebration, and my happiness bubbled out, communicating that attitude to my school family. People looked at me in awe of the miracle of healing that was taking place in my life.

As I returned to my room after meditating under the huge eucalyptus trees on campus, I was startled by a 3½-foot-long snake, which stirred ever so slightly directly under my foot. I almost lost my balance from having to move my right foot to avoid stepping on her. The snake did not move much when I yelled, evidently less frightened of me than I was of her, so after I regained my equilibrium, I examined this real live symbol of healing more closely. I let go of my fear of harm and tried to make a relationship with this magnificent creature, which had slithered out of my dream into my waking world. She also watched me as I examined her long racing stripes bordering a central stripe pattern of horizontal dark and light bars, and we became friends.

September 21, 1991, Journal:

Peace floods my soul. Sitting alone with my journal at the end of a long day of study feels good.

Inside, the question of whether I really want to live or to die has been firmly settled: I wish to *live*. However, I am also not attached to that wish. If dying were in my destiny in the near future, I could accept that option.

On September 25, the day after I returned from California, I had the first radiation treatment. Soli met me at the hospital, and I squeezed the initial treatment into my work schedule following a home visit in the area, cutting a little time out of the visit, telling my client only that I had to go to "a meeting" that came up suddenly at the hospital.

I felt a bit flighty, but also curious about the technology involved and glad to finally get underway with the treatment. Tim, the radiation technologist, was friendly and explained the procedure, taking lots of measurements to check out everything and shooting another set of X-rays to be sure the calibrations were correct. I had to lie perfectly still on my back on the table with my arm raised above my head for half an hour. My arm fell asleep, but this time it did not hurt nearly as bad as it had when I had to lie perfectly still in that uncomfortable position for an hour for the marking.

The actual treatment consisted of two radiation shots of only 45 seconds each, one from a far left angle and one from the far right. A red laser beam ran across my breast to line up the machine, which made a high humming noise when it was on. Strange green lights flashed on and made funny noises when the X-rays were taken, and I felt as if I were in "Star Wars."

Tim said that nobody but nobody could feel X-rays on their body. During the radiation blasts, I envisioned healing rays of sunlight coming out of the palms of Lord Jesus Christ and going into my breast. I visualized healing energy in my breast and felt a pleasant tingle.

Afterwards, I tried on two of the free wigs the radiation department gave out. Soli commented on how old they made me look, and the mirror confirmed his observation. The lines of my face seemed severe and aged without my real hair framing them, no matter which color I tried on. I felt ashamed of how old I appeared in the mirror beside Soli, who looked so youthful. I did not want to look as if I were his mother instead of his wife!

That night my breast hurt while I was resting before supper, and when I felt it, I was shocked and aghast at the change in its texture. My left breast, which had always been slightly smaller than the right side, was swollen and felt much firmer, larger, and warmer than its mate. It stood up by itself, and its skin texture was irregular because of the swelling from the bulging veins on the bottom side, which already had stretch marks from my breast-feeding days. With the magic marker dots the hospital had used to enlarge the tattoo points so they could see them to aim their machines right, the whole appearance looked pretty silly.

Tim had warned me that my breast might swell and move the tattoo points a bit, but I had not expected the results so fast. Cancer and radiation were both mysterious and unseen, yet with sudden and drastic effects.

Soon I had another snake dream:

September 27, 1991, Dream:

Snakes are around. I correctly identify the copperhead on the left. I am unsure whether the one in the center is a harmless gopher snake or another kind. A third one is on the right, unidentified.

As I am about to go up the steps to my house, I hear a rustling in the weeds and know a snake is there. What kind? I am wearing no underwear and am afraid that the snake may bite my vagina if I go up the steps. I think I see it coming directly towards me. The thought of the snake as healer flits through my mind, but my emotion is just plain *fear*.

My battle with fear was not over: it had just entered another round of the struggle. The fear of the snake biting my vagina is a blatant example of fear of the phallic nature of the serpent, a fear linked to my early sexual abuse, a wound that had not yet healed.

> The presence of a serpent is almost universally associated with pregnancy. It accompanies all female dieties and the Great Mother, and is often depicted twining round them or held in their hands. (Cooper, 1978, p. 146)

While my physical body was headed for menopause instead of pregnancy, new psychological life was incubating in me. For two years, I had been having repeated dreams that I was pregnant and about to have a baby; however, my fear blocked the process of spiritual birth within me. My sexual wounds and my fear of men were holding back my psychological development.

The physical toll of the treatments made me tired and lowered my sexual libido, shaking up my sexual identity and frightening me with the thought that I might not be able to function sexually in my role as Soli's wife. By the weekend, after the third treatment, depression had rolled into my body full force:

September 28, 1991, Journal:

Today when I finished my balanced breathing exercise and began to meditate, instantly a torrent of tears broke loose.

I cried for a long time. Depression sits on me again, and I am drawn to thoughts of going home (dying). Part of me really wants to leave the earth plane, and the other 80% of me wants desperately to stay, especially for Robin, Vivien, and Soli. In prayer I asked for divine permission to live out my time here to fulfill my work.

I had terminal insomnia last night, awakening around 4:00 A.M. and not getting back to a light sleep till around 6:30. Is the radiation treatment putting me into this depression?

I talked with Vivien and told her how I felt. My insides are fragile. Being outside today planting daffodil bulbs all afternoon felt good; however, I got physically worn out and did not have enough energy to finish the job. I was glad for Soli's help to complete the task.

My body has felt so bad that I have not wanted to have sex for the past week, and I sense that Soli gets grumpy and a bit argumentative when sex is missing from our relationship. I talked with him tonight to let him know how I am feeling. I cried in his arms for a long time.

I asked him how he felt when I said that the desire to "go home" was strong inside me. He first answered with some abstractions I did not understand, then said that life and death are not the only two options available to me. I disagreed and said that with cancer, the choices were plainly either living or dying. He then agreed, and when I asked him a second time how he felt about hearing me talk about my death, he said he felt compassion.

September 29, 1991, Meditation:

I realize that time spent in meditation is "home time," my chance to be at one with my spiritual roots.

Tears burst out from the pain in my body and my soul, and I see the image of a cross where my body is. I feel the crucifixion process within me.

Then the inner figures form a ball of light in their palms, about five inches in diameter. They implant this light into my left breast to protect my body and to heal the radiation trauma. I am grateful!

At the end, I consecrate my body again to my divine source. If my greater service is to live, I will do so with joy. I also accept the possibility of death and realize with a bit of surprise that death really is my friend, not an enemy to be feared, but a friend to guide me back home when my time here is finished.

In my heart Lord Jesus Christ says to me, "You are my beloved daughter. You came from me, and some day you will return to me."

I pose a mental question about a more exact return date and I hear, "Prepare for a long time yet. You have much work to do."

CHAPTER 11

New Vision

The most frightening aspect of the depression that came with the radiation treatments was its lowering my will to live. I feared that if my desire were to die (like Marcela's), my life force would not fully energize my immune system, and my body would have more difficulty fighting any cancer cells that might be present. With this new wave of depression, some of the ambivalence I had felt towards Soli as a newlywed returned, as we clashed over financial issues.

Before the radiation treatments began, Soli and I had moments of sexual ecstasy:

> *August 28, 1991, Journal:*
>
> After making love with Soli, as we lay with his chest pounding on top of mine, I felt a physical opening between our hearts, with energy flowing between them. Wonderful! As we cuddled in spoon position afterwards, I experienced bliss. I felt myself falling in love with him again and felt the triangle that the two of us formed with God.
>
> I thought, "I am lying here close to the man I love, soaking up the love connection with every skin cell that is touching his body." Deep contentment filled my soul.
>
> I thought about how happy my neuropeptides were and how wonderful my Life was, about hearing Bernie Siegel's talk on the healing power of love the previous night, and

> about how this feeling of close connection to the man I love was what I had been wanting all my life.

But as the depression wiped out my sexual energy, in moments when Soli tried to discourage me from spending so much money on nutritional supplements, the pendulum of my feelings for him swung from appreciation to irritation. I have a Sagittarian nature, which is portrayed in astrology as the centaur riding freely across the plain, and I felt crippled by the limitations of the cancer and the boundaries my husband was trying to set around my spending. I felt vexed by the fence Soli was attempting to erect around me. I did not want to feel controlled by his values, which differed from mine, since his needs were quite simple, and my needs seemed complex by comparison.

I was fighting at a deep level to find the will to live out my time here, battling the depression that seduced me towards death, and I felt that vitamins and herbs were absolutely necessary to repair my body so I could get through the medical treatments for cancer.

One of the first things I did after surgery was to go through my catalog of nutritional supplements and order one of everything that I thought might help me. I had been health conscious even before I discovered I had cancer and had already been taking a multiple vitamin and mineral supplement, extra Vitamin C, extra calcium and magnesium, the rejuvenating herb gotu kola, another herbal energizer, and the general tonics taheebo tea and aloe vera juice. After finding cancer, I added to my list a supplement to stimulate the adrenal glands for heightened immune system functioning, odorless garlic pills for digestion, an herbal blend for intestinal health, homeopathic shark cartilage to promote healing, super blue-green algae for nutrition, coenzyme Q-10 (mysterious in function, but the book said it was essential to the life and health of every cell), both Vitamin E and a water catalyst for anti-oxidant properties, and capsicum to enhance the properties of the other herbs.

Soli feared that taking all of the vitamins and herbs in combination might have some negative side effects and thought the synergistic effect would lower the requirements for each individual item. He was also concerned about the financial stress on my budget of paying for it all, since these supplements came to $95 per month plus another

$95 per month for the herbs prescribed by the Ayurveda center. Soli felt confident that I could heal without so many extra herbs, pills, and healing treatments and that the financial distress they produced might outweigh their medicinal value.

After 38 years of being a bachelor and accumulating savings because he spent very carefully, the financial jolt of these unexpected expenses from his brand new family was hard for Soli to handle. On the other hand, I wanted to be as healthy as possible under the circumstances and felt that my task was to nourish myself the best I could. Besides, even with three college tuitions, Soli still had quite a bit of money in the bank, and I was paying for my vitamins and herbs out of my own paycheck. (However, if Soli had not paid off one of the mortgages on our house when we got married, I would not have had enough money for all of those supplements, and I could not have afforded to cut back on my time at the clinic.) The price for what I felt I needed to preserve my life seemed small to me.

When I wanted to contact Gloria Karpinski,[1] my spiritual mentor, to help me better understand the process I was going through, Soli again questioned the added expense. However, I contacted her anyway because I knew my soul needed the loving guidance her insight had given me in the past, and I desperately wanted to make sense out of the chaos within.

The burning question inside of me throughout the first few weeks of knowing I had cancer was *why* and *how* the cancer had entered my life, and I asked my mentor these questions as well as what the inner meaning of the cancer was and how Soli and I could improve our relationship.

I got together with her just after the Kirlian box (see Chapter 14 and Appendices B & C) was finally working. In prayer, my mentor attuned to me spiritually, mentally, emotionally, and physically, and then talked with me about the symbols she saw. She called this process an "attunement." Gloria encouraged me to weigh everything she reported very carefully and reminded me that the truth about who and what I am and about the meaning of this experience was already inside of me.

The images she brought forth were rich and resonated with truth, moving me deeply and giving me a new vision of the spiritual

---

[1]Gloria is the author of *Where Two Worlds Touch* (New York: Ballantine Fawcett Tradebooks, 1990).

purpose of my life and quieting my fears of death. I especially loved the stories about past lives[2] with Soli, regardless of whether they were actual lifetimes or metaphors from the collective pool of our ancestors about the qualities in my relationship with Soli during this life.

I saw that although my journey seemed intensely personal, it was intricately connected to the matrix of the hearts of all women, and the initiation was about much more than just surviving the physical illness of cancer. Because of the depth of wisdom of the attunement, I could only begin to grasp its meaning in my life, and I am in an ongoing process of working on the issues it brought to light.

Attunement by Gloria Karpinski
for Barbara Stone
October 2, 1991:

Much of the answer you are seeking about the meaning of this experience has already been given to you when your own vision said, "This is not about death, this is about Life." But when I looked at that statement, I saw that to enter into life, a death has to come first. All Holy Scriptures remind us that a reality of the spiritual path is that passages come of dying to whatever we are attached to:

China: "The student learns by daily loss."
Krishna in India: "The self dies that the self can be born."
Jesus: "You have to lose your life in order to gain it."

These statements are references to destruction of what we call the ego. You are doing a form of dying in order that birthing can emerge, like the phoenix rising from the ashes, and the passage is necessary. You could have picked many

[2]For further reading about the theory of past lives, see *Many Lives, Many Masters* (Weiss, 1988), whose book cover says it is "The true story of a prominent psychiatrist, his young patient, and the past-life therapy that changed both their lives" or the book by Jungian analyst Roger L. Woolger (1987), titled *Other Lives, Other Selves.* In addition, a film of a Past Life Workshop by Roger Woolger and his wife Jennifer is available from The Hartley Film Foundation, Inc. (1988), Cos Cob, CT, in the series "Films for a New Age."

other ways of dying, but when this dark, chaotic passage of embracing your worst fears is over, you will be invincible. *And it will be over, because I am very clear it is a passage and not a conclusion.*

Once you have embraced the death of whatever your ego thinks is the most important identification, what can anybody possibly threaten you with?

She said that I was at the edge of a dark passage in which the light would have to come from within myself, from all that I had ever learned, and that the terror I was feeling would give way to peace and an inner sense of knowing truth.

This is then a major initiation. It is taking you through a walk in which you are going to sacrifice to God whatever you thought was important, whatever Barbara Stone's ego would have clung to, to say, "Oh this is who I am, I've got it now. I am Soli's wife, that's who I am. Oh no, no, I am a Ph.D. candidate, that's who I am. I am the mother of Robin and Vivien."

The Spirit says to lay down whatever the ego has been clinging to when you reach these levels of initiation.

Gloria said that, like the Bible story of Abraham being told to sacrifice his only son Isaac, the point was not that God wanted me to die to the things I thought were precious, but that I had to be willing to make the sacrifice.

You can face this passage, including the radiation, including the threat of losing your life, including the threat that now at last you have found Soli and here he could be ripped away from you, including suffering chemo; and either become angry, bitter, terrified, really put out with God, feeling separate and abandoned; or, let the process become a glorious sacrifice to God.

God wants everything in you, absolutely everything.

She then addressed the question of why the initiation came in the physical body since an initiation could be faced in many different ways.

I see a deep seated and very subtle ambivalence about being on the earth plane in the first place, the kind of attitude you can get away with as long as you don't try to become an initiate. As long as you don't try to become conscious, you can drink alcohol and eat red meat, and get away with it, living to be 90 years old and being happy and healthy as a horse. But the minute you say you want to get conscious, all of a sudden those things are no longer compatible.

With most people, the first time trouble starts is when they say they want to get conscious. God takes them very seriously, and all of a sudden their life becomes chaos, because everything that isn't in line with the program has to start going.

The ambivalence she talked about had not been a problem at my former level of functioning, but I was moving into a different vibration, and at this new level, this "hidden enemy" would compromise my mission. The three components of that ambivalence she identified were *despair* at the violence present in this world, *abandonment,* in which I sometimes felt alienated from any divine connection, and plain old *fear* of what organized groups will do, like a mob or an army, and particularly if those groups were male.

She felt that these three elements had been squashed into what Jungians would call "the shadow" so I would not have to transform them and had been wrapped up in a package like "a black inky thing" and hidden in a corner of my psychological basement. Then every time someone in collective society went through an abandonment issue, that ripple activated that piece of my shadow so it could come up to get purified.

> That's the point now, both individually and collectively; *nobody has hiding places anymore.*
>
> The focus of your cancer was in breast tissue. A lot of your fear was of the other, which is men, what they do to women, of the abuse of not only females, but the Divine Mother, all the energy of despair that is held by women all over the world that is now surfacing in the present escalation of brutality towards women, which is the backlash against the emerging Mother, fighting evolution.[3]

Gloria added that going through the process of cancer with consciousness would affect the feminine principle everywhere, transforming the fear and despair my cancer represented into new life. Once I let my old patterns die, I could turn within to find my answers instead of looking at outer things like a degree, money, or marriage as "the solution."

> The first thing in the dying that I am speaking of is dying to any sense of a tomorrow. The power that high initiates have is in their knowing that everything is in the moment.

[3]Brutality against a woman often comes internally, from an inner male figure that abuses the woman by putting the idea into her head that she is worthless or does not deserve to be treated well. Likewise, the inner feminine in a man gets brutalized by male values in society, which say that a man must be "strong" at all times by showing no emotion, denying his feelings. The battle is not between the sexes, with all women as victims and all men as rapists. It is between the feminine principle and the masculine principle, the yin and the yang in Chinese philosophy. These two polarities need to learn to work together in harmony instead of fighting each other.

They have everything at their command, because nothing is clinging to the past or practicing for the future; it's just in the moment. The threat of cancer brings you right up against the moment, rearranging all your priorities.

Before you are through with this process, you will look at Robin and Vivien differently. Instead of saying, "Oh my God, maybe I won't be around to see their children," once you face and embrace that fear and actually live with it, it's like, "Well, maybe I won't."

Then the fear is replaced by thanksgiving that you were able to share this part of the life walk with them. And that thanksgiving becomes a whole different way of living, because then you are continually grateful for the next moment, instead of looking at what is denied.

The empowerment in the attitude of living in the moment is almost impossible to estimate. One way to immediately grasp this point is that an orgasm cannot be enjoyed if you are thinking about tomorrow. You have to be in the moment, totally present with everything happening, to experience anything ecstatic.

She encouraged me to create an imagery of "the black inky thing" in my personal therapy and to give it a voice, because as it spoke, it would lose energy from the love and attention I gave to it. She recommended embracing this energy with a therapist present so I would feel safer during the process of confronting these feelings I had so long denied.

She said that completing the initiation, eliminating all the feelings of despair and separation from God, would take seven years to complete, though I would not have cancer for that long. And though I was presently in the preparation stage, where the suffering comes, the initiation itself would be liberation and would bring "mastership over the body."

One of the next steps that we are going to actually go through in evolution is the literal raising of the vibratory rate of the human body, one reason why we have so much obsession with physical fitness around the world. We are on the edge of this vibratory shift, and so we have a lot of Kundalini action going on. Part of what activated your cancer was Kundalini energy, which brings me around to Soli.

The sexual energy with Soli has been a trigger point for the cancer. Unlike other sexual experiences, Soli has activated the energy in all your bodies. His interaction with you has mated with you not just as a physical woman; he has also mated with you etherically, emotionally, mentally, and spiritually. Hear the difference? Mating at all levels has caused a tremendous activation of Kundalini energy, and when Kundalini goes, she activates whatever is latent.

Gloria mentioned that she noticed a little bit of disregard for my physical body in me, that I wished I could just zip around without it to conduct my mission on this earth.

But when this passage is over, you are going to *love* your physical body, and you will have claimed your physical vehicle, so you will not be half in your body and half out. You will really value it.

I see that the cancer is contained, just in the area that is being treated. I would be very surprised to hear from you that it spread anywhere. You never planned to lose the physical body over this illness. You're not going to get off that easy.

Gloria felt that I had attracted Soli into my life for this passage because of the fit between us mentally, physically, emotionally, etherically, and spiritually.

Soli is himself a very developed soul who agreed to come into your life to be with you at this juncture to be the support that you could stand on. He is an underpinning for this passage.

The lifetimes you have shared with Soli are like pearls. I see multiple stones, crystals, sapphires, and pearls that are highly symbolic of the nature of your relationship with each other.

I loved the stories Gloria told about these past lives. She saw an image of a huge book with the soul names of Soli and me on the cover. Each page was a lifetime we had shared, from the times of the early Christians to Napoleon to Native Americans. She saw themes emerging of successful partnership, psychic rapport, and spiritual courage.

Clearly you and Soli have come together for multiple reasons. Both of you are going through major initiations in this lifetime. Soli is working on a lot of synthesis of mind-body-spirit. To be any help to you, he has to surrender control, but he has leadership ability, a strong core I feel in his energy that I really like. Soli doesn't disappear when times get hard.

I see the two of you are also together to provide an eyeball-to-eyeball relationship, a marriage of equals. As you go deeply into your forties, the two of you will be role models of a new form of marriage. Evolution is now challenging our ideas about what is a man, a woman, a marriage. All of our shadows in these areas are up for purification.

Because you and Soli are both powerful old souls with tremendous history behind you, both together and separately, the two of you have come together now when you are through with a lot of your personal karma for mutual support of your missions, to empower each other through your initiations.

She encouraged us to connect at the spiritual level to get past the conflict at the mental level over which way we were going to cut up the green peppers for supper or how many vitamins to buy.

> The transformation you are going through at the cellular level by confronting and moving through the terror of death will give an authenticity when you speak about healing. You will work at the interfaces between a person's psychology and their spirit and how that mind-spirit connection effectively transforms the physical body.
>
> As we speak, I see standing behind you the figure of Kwan Yin, with you being a light resting in the rays of energy that emanate from her heart, like being fed from the breast of the Divine Mother, the cosmic breast, nourishment that doesn't leave you hungry, being loved from the center of the heart, unconditional and ongoing.
>
> You have to be willing to go through this passage and find that light within yourself. In the worst moments of the fear, she is there. Although the human ego has to go through the appearance of being alone, in fact you are never alone; you are in her radiation the entire time.

The attunement filled me with peace and joy. Upon first hearing it, the parts that stood out were the ones that said that I was not going to lose my physical body to the cancer, easing my fears of recurrence of the illness. The attunement also enhanced my respect for Soli, as I saw him in a new light, realizing that beneath our surface conflicts over financial issues, we were ideal partners for each other because of the deep spiritual bond between us.

Instead of feeling sorry for Soli because when he finally got married, his wife turned out to have a major health flaw, I saw his soul had chosen to come into my life to set the stage so we could both work on the issues necessary to get through our respective initiations. I began to focus on the positive aspects of my relationship with Soli and to appreciate his strengths more, seeing the areas of conflict between us as opportunities for growth.

The attunement opened up a deeper level of understanding of who I am and the nature of the initiation process itself. I saw that the internal agony of this initiation was similar to the pain I had gone through 10 years earlier when I was going through the initiation fire of divorce; the words came to mind of a song I had written out of that pain and reflected the same despair Gloria talked about of the ego being called to give up its attachments:

OH MY LORD

Oh my Lord, Oh my Lord,
I cry to You.

Oh my Lord, Oh my Lord,
I worship You.

Pain opens my eyes.
Pain opens my ears.
And I grow more open through the years.

If my life were my own
And death were the end,
I wouldn't go on, no how.
But death is a door
And my life is yours.
So where am I going now?

Some folks say that you have a plan:
Whatever gets between you and a woman,
You take away,
So she can listen better on the following day.

Are you doing that to me now?
Is that why everything that I hold dear
Suddenly seems to disappear?
Where do I go from here?

Oh my Lord, Oh my Lord,
I cry to you.

Oh my Lord, Oh my Lord,
I worship You.

Pain opens my eyes.
Pain opens my ears.
And I grow more open through the years.

I realized that this heart-wrenching process of having everything my ego was attached to placed on the sacrifice altar was inherent in the initiation process itself and necessary for purification.

I redefined my cancer as an opportunity to develop mastership over the physical body instead of viewing it as a mistake in my DNA. The gift of the attunement harmonized my emotions in myriad subtle ways and put into my heart an unconditionally positive attitude towards healing from the cancer, a mental state that may encourage strong immune system functioning.

CHAPTER 12

Ups and Downs

The first radiation treatments had plunged me down into the pits of despair, but the attunement pulled me back up again by giving me a vision that not only would I survive the cancer, but that I would grow tremendously from releasing my fear and despair and learning to focus all the energy of my spirit-mind-body on my own healing.

The radiation had put a heavy drain on my physical energy, and my mood continued to roller coaster throughout the two months of radiating the entire breast thirty-three times and the local area of the tumor site an additional seven times.

I got the 15-minute radiation treatments daily Mondays through Fridays at 8:15 A.M. so I could get to work by 9:00. I felt fatigued all day every time I had a treatment, dragging myself around at the clinic and unable to stay awake in the afternoon without a nap during my lunch hour. Not having enough physical energy to work at my former pace frustrated me, as I still had the same intense inner drive as before, but not enough energy to accomplish all I wanted to get done.

Although staying awake and alert with clients was physically difficult, during the therapy sessions, I felt deeper insight moving into my work with my clients and into understanding my own psychology:

> *October 5, 1991, Journal:*
>
> Suddenly I understand that the reason I used to resist going to bed at night and used to have to force myself to

meditate is because both sleep and meditation resemble death in that my conscious ego is not in control.

Now that I am making a conscious decision to embrace my death, I look forward to the time when I can go back into the arms of God. Since I no longer fear the change involved with death, I have also lost my resistance to sleep and meditation.

Meditating today I felt myself very much in the presence of God, "home time." I felt a flexible steel rod being installed through my spine with one end deeply grounded in the earth and the other end in the heavens. Energy is channeled to me through this rod. When I am weary, I can lean on this rod to hold me up. Though it has the capacity to bend infinitely, it will never break.

My left arm had basically healed and was restored to full range of motion by one month after the surgery. By two months after surgery, the scars from the two incisions were fading, and the bunched up skin from the staples under my armpit returned to lying flat. The black and blue mark around the stab wound for the drainage tube was also completely gone.

By the time I had completed ten of the radiation treatments, though both breasts would still fit into a size B cup bra, the left breast was swollen to about one centimeter bigger all around than the right breast, and the nipple area was proportionally more swollen than the rest. Though the skin looked slightly sunburned, it did not hurt.

The radiation oncology team at the hospital was sensitive and emotionally supportive, and since we saw each other every day, we became close. I was touched by Tim's story of being a radiation technologist for seven years, since the job usually burns people out after just two years. Tim wore an insulin pump for his diabetes and checked his blood four times a day, reporting that some people wonder how he can put up with all that; but he felt the routine was easy when the alternatives were considered, adding, "Nobody can take their life for granted. You just have to enjoy each moment as it comes."

On the four days a week when I did not work, I spent time in meditation, took long walks in the woods, fed myself nourishing food, and listened to self healing tapes by Louise Hay, Bernie Siegel, and Emmett Miller while resting on the couch.

One of the visualization tapes I worked with asked, "What holds you back from giving your symptom what it needs?" I saw clearly that fear was holding me back and worked to release that fear. As I let it go, in my mind's eye I saw beautiful wolf heads, and the following visualization ensued:

October 14, 1991, Visualization:

To help me, my inner spiritual advisor gave me the spirit of the wolf and asked me to ground the insight:

The wolf spirit in me leads me to be a better animal:

Rest when tired.
Go after the meat when needed.
Fight for territory.
Sniff out danger.

In short, the visualization encouraged me to attend to my needs.

I was able to handle the effects of the radiation by devoting time to my needs for rest and nourishment until I got knocked off balance by conflict that erupted between Soli and Vivien over housekeeping issues.

October 18, 1991, Journal:

I feel sick emotionally and physically. The radiation treatments are taking a big toll on me. I have sixteen down and ten more to go on the entire breast.

I'm glad my doctor is giving me a week's break in the treatment. I need a rest. My skin is red and it itches terribly in a three and a half inch circle on my chest. The whole breast is red and aches. Also, I need some energy for going out to California.

Last night's blowup between Soli and Vivien hit me hard, at a time when my physical energy was already very low. I couldn't stand Soli giving Vivien direct orders about cleaning her room. He was trying to practice the concept he learned at family therapy this week that there is a time for authority, but I did not like the way he was directing anger at my daughter.

I was attempting to recuperate on the couch, listening to a healing tape, but I felt Vivien was getting dumped on and couldn't stay out of it. Then Soli fell apart emotionally when I confronted him so directly, and he felt lost. I felt horrible and wanted to get away from them both and decided to go to bed early, at 7:45 P.M. The argument took away my appetite. I felt so pissed because I didn't get what I needed: peace and harmony to recuperate.

The whole incident threw me into a depression state again, which I couldn't shake. This morning Soli tried to cheer me up by suggesting I have an AIDS test to see whether it's really worth going through all this treatment. *(His idea didn't help!)*

To try to cheer myself up, I stopped by the store on my way home from the radiation treatment today and bought a beautiful cotton and linen black suit, a pair of purple stretch pants, a purple turtleneck, a teal shirt, and three pairs of underwear, all for $60.

Crying a lot this afternoon lightened up the depression (see Kirlian photographs 2 and 3, Chapter 14) and meditating helped (see Kirlian photograph 4), but when Soli got upset tonight about the money I spent on new clothes, I got off balance again. When he is off balance, I feel drained too. And right now I feel physically sick.

The disagreements over how clean Vivien was supposed to keep her room were a normal part of working out expectations in a new family, especially in our tight housing situation where the three of us shared a small house of only 660 square feet. However, in my weakened state, I could not handle even ordinary, routine stress.

The oncology nurse told me that people bounce right back after the radiation treatment, comforting me at a critical time; I could not have endured the emotional pain involved in this passage without knowing that it would end soon.

The next day as I sang and chanted prayers, the following visualization came:

October 19, 1991, Visualization:

Lord Jesus Christ comes saying, "I bless you."

Tears break loose in me because of my pain, and *I think this blessing feels like hell.* I feel the suffering within. After I cry, Lord Jesus Christ says to me, "Give me your fear. Give it to me."

I image the black box down in my cellar and see that it is indeed filled with fear. With more tears, I give it over to Lord Jesus Christ. He/she holds it in his/her hands and exposes it to Divine Light, which permeates it. Lord Jesus Christ says it is energy, but in negative form in fear, and washes away the negativity.

He/she says, "It is only fear of getting closer to me."

After the murky mess is cleansed, Lord Jesus Christ says again, "I now heal your soul."

I feel a sensation like a spiral vacuuming out my heart chakra, and I feel negative energy being cleaned out of my chest. It lasts about a minute or two. I feel lighter and am told, "Go in peace."

That afternoon I got another treatment by Marie (see Kirlian photographs 5 and 6), which felt great. By two days later, on my "vacation" from radiation, my energy had bounced back just like the nurse said it would:

October 21, 1991, Journal:

Today I feel I could move mountains. Physically I feel much better, and the irritability I felt so strongly yesterday all day (I felt like a grouch) has dissipated.

I feel as if I am better than before the whole cancer came. But the energy within feels like wild horses trying to burst loose, and I border on feeling like a nervous wreck.

I still want to do so much, read everything, write everything, make jewelry, write letters to everyone, and spend time on my inner healing. There are not enough hours in the day to accomplish all these things.

Yesterday I felt my relationship with Soli healing up during our walk in the woods. I got more grounded in who I am, more certain that the cancer is the precipitating factor to push me up into a higher state of functioning.

My friends Ray and Starr stopped by tonight and did a group healing circle with the three of us. When they came, I said, "I have all this energy and I feel wired, as if I could light up a light bulb with my bare hands." After praying together and repeating the key phrase, "I accept and share the alignment of my mind with the Divine mind," I now feel centered and balanced. Peace and joy reign inside me again.

My emotional seesaw continued to pivot as Soli and Vivien kept on arguing. I wanted to be close to both of them, but also to stay out of the crossfire as they worked out their boundaries and expectations. However, in my weakened state, I could not tolerate the drain the disagreements between them put on my emotional energy, and I felt like moving out.

October 24, 1991, Journal: (on the flight to LA for my doctoral program classes)

A sadness, a deep sadness is within me. The past few days have been difficult, and a fight is going on within me between my fear of dying and the depression, which wants me to die.

Soli was supportive this morning, recognizing his role in the drain on my energy and wanting to do better in the future.

I have come to face my own limitations: I can handle physical illness and stress at work both, but not simultaneous tension in my home life. Then *there is no safe place for me.*

During the weekend in California, when I got some distance from the stressors at home, I talked with my school friends and came to realize that both teenage daughters and stepfamilies are inherently difficult, and when the two are mixed together with a little cancer thrown into the works, the result is bound to be complicated. A dream came to shed additional light on the difficulty of living in a stepfamily, an emotional reality that Soli, Vivien, and I were all facing and not dealing with very effectively. We had not had time to get our roles clarified before the added stress of cancer invaded our family.

Vivien was suffering from feeling left out because of my union with Soli, and I was too preoccupied with cancer to notice her feelings.

October 26, 1991, Dream:

At my sister's house, I am gathering up my things. She is preoccupied with what she is doing and does not talk to me. I am missing my wallet and ask if anyone has seen it. My stepmother says she took it because it had pictures inside that were too sexy and were anti-Jesus by her standards.

I take my wallet back out of her purse. She says it has her husband's paycheck inside, and I give that back to her.

Overwhelmed with feeling, I tell her, *"Stepfamilies are hard."*

I say that I am twenty-four years old, an adult, and if she will not treat me as an adult, I will move out. Also, if she wants me to move out, I will go. I feel distanced from all the others in my family.

At the very time when we most needed one another's support, all three of us in the family felt distanced from one another, preoccupied with our individual worries as our values clashed, primarily around issues of housekeeping, stepparent rights, privacy, and Vivien's struggle to move from child status into adult status, all themes reflected in the dream.

Another dream highlighted the emotional turmoil of the cancer treatment process. The beginning of the dream showed my blatant fear that I might not survive the illness, and the dream then progressed to images of surgery (being poked with sharp tools) and radiation (being heated up almost too hot to survive and being shocked).

October 29, 1991, Dream:

I am crying about the cancer in my left breast. I feel it, hooking my finger under the bone. It is about the size of a walnut. I ache from the pain of it inside me and feel unsure whether I will survive it.

A black guy on the street slams into me. Then he is white and wants to poke me with sharp little tools. I want to tell him I have cancer and am getting enough physical abuse for the moment.

A baby is inside a bottle. I am concerned that it got too hot, but the baby comes out okay.

I get a repeated shock to my arm from a cord dangling from the ceiling. I go upstairs to have the owner fix it. When she hears of the problem, she does a handstand flip out the second-story window and lands smoothly on her feet on the ground. She goes around to fix the problem.

I carefully go out the window onto the garage roof. Then I fall down and panic. I grab for something to hold onto in order to break my fall and get something that slows me, and I lose my fear.

Just at that moment in real life, Soli woke me up by yelling in his sleep. I asked what he had been dreaming, and he reported that in his dream he was going across the street to make a phone call, and in the dark, someone grabbed him. We both laughed for a long time.

In the dream, I was in a state of panic and grabbed something that slowed me down and quieted my fear. That something I grabbed was Soli. His moderate pace slowed down my internal race horses, which were chomping at the bit, and the sense of protection he built around me helped me deal with my fears.

The next night I dreamed that Soli and I were shopping together in the Soviet Union, and I was considering getting a white sweatshirt with a huge yellow and red snake plastered across it, a symbol for the healing process going on within me as I walked through the foreign land of illness with Soli at my side.

Another dream pointed out a faulty internal attitude that needed correcting:

November 4, 1991, Dream:

The house I lived in as a child is a great mess except for one room in the basement, where my sister and brother-in-law have made a nice breakfast for some of my relatives. I have eaten half my food and hear the guests arriving when I realize I have on only a shirt and underpants and go upstairs to find some slacks. The only clothes I find in the closets are for other people, and I weep, "Why is there room in my home only for *other people's* clothing?"

I want to send word to them downstairs to save the rest of my breakfast. I want to eat it.

Changing an ingrained pattern of having room only for other people's needs is not easy, and the dream was reminding me of the continuing importance of putting my own needs on my list, making some room in the closet for my clothes too, and eating all of my breakfast before I run off to take care of everybody else.

During this time, the dream that kept popping out at the beginning of my dream file on my computer every time I made another journal entry stated, "The people I've been helping don't need that much help. My energy is needed elsewhere." This dream message was a constant reminder to examine my pattern of nurturing, as the place my energy was needed was right inside my own body.

Physically nourishing myself during radiation treatment was difficult because I had diarrhea throughout the entire treatment time. My weight kept slipping down below normal. Never in my entire life had I had the problem of having to *try* to put on a few pounds. I had always struggled to get a few more pounds off.

November 4, 1991, Journal:

The radiation oncologist surprised me today when he said my diarrhea (quite active the past few days) was unconnected to the radiation treatments because they were treating the breast, not the abdomen. That's like saying the parts of my body are not connected! That was a totally new thought.

The doctor also prescribed a steroid cream to apply to the treatment area three times a day, triamcinolone acetonide. The skin doesn't hurt, but it looks pretty red. Could I have lost all feeling in it?

On the phone tonight, Dad said, "Your letter made us feel depressed—hearing how your skin got all red and you were having trouble." I told him that the cranial treatments that Marie has continued to give me are helping and that I feel super lucky to have Marie around.

I see protrusions on the lower half of my breast in rays where the blood vessels have swollen up. I wonder if that effect will be permanent. The area right around the incision is very dark, looking almost freckled, and the nipple has three colors: normal light brown, dark brown patches, and spots with almost no pigment at all. Still, it's worth trying to hang onto my breast.

I feel concern about a couple of tiny lumps I feel in my right breast. I will have Dr. Orr check them when I see him this month. The thought of having another surgery or radiation treatment battery is overwhelming to me at the moment.

My dreams continued to guide me, easing my fears of a recurrence of cancer. One very clear dream pointed out the need to totally redo my concept of the female side of God, the Great Goddess, instead of just patching up my image of Her:

November 6, 1991, Dream:

I walk around school with my chest fairly swollen. Someone inquires what is wrong with me, and I say, "I have breast cancer." I realize it could be worse.

During a visit with Ray and Starr, I realize I need to make a complete new goddess statue rather than trying to mend the one with a broken arm. They say the lump in my right breast is cancer, but the chemotherapy and the new goddess figure will fix it.

November 7, 1991, Journal:

Only two more treatments to the whole breast! The excitement about ending this ordeal, which feels as if it knocks out my electrical system, is almost enough to make me feel good again. I notice progressive fatigue as the treatment effects build up.

For the last week I have been crazy about wearing purple. I love it! And I feel good in it. My supervisor at work and others commented on how perky I look and praised the energy I have been able to bring to the health center.

When I finish this passage, I will be very grateful to have my body restored to feeling good again. I have brief times now when I feel well, but after a day at the clinic, my physical energy feels drained. Listening to the horror stories about AIDS, crime, and substance abuse today brought out feelings of deep sadness within me.

I needed to move into a less stressful job situation for the sake of my own health, but I could not yet bring myself to turning my back on the needs of my clients so I could attend to my own needs. I still had mostly other people's clothes in my closet.

My physical and emotional state reached another low point just before the final treatment to the entire breast:

November 8, 1991, Journal:

I feel terrible. Sobs kept welling up as I listened to a Bernie Siegel tape this afternoon. My physical body feels very un-at-ease. I have a slight headache and intestinal discomfort and a general feeling that something within is very wrong.

Today when Tim mentioned radiation being a contaminant to the body, I once again felt my body's invasion by the poisonous substance being used to burn out the cancer.

I carry my breast as a wounded member, and I am amazed that the breast itself seems to have found a way to cope with the onslaught of radiation. No longer is it as hot as it used to be. The swelling has gone down slightly, even though the treatments continue, and the steroid cream has healed up the skin discomfort. Once in awhile a spot will itch, but the skin usually feels good. I can even wear a necklace again at times.

This evening we went over to Marie's for dinner and a cranial manipulation treatment. Marie felt ill herself before beginning the treatment but agreed to try working on me to see what happened. As she began by holding my ankles, she commented, "Radiation is taking it's toll. You feel like an ordinary person right now."

We all laughed a great deal, and I let her know, "Nobody ever accused me of being ordinary before!" She worked on me longer than usual, feeling the energy blockages in my system.

As I lay on her table being worked on, I asked Jesus to have mercy on me. I felt the accumulation of the effects of the radiation, and the full extent of the damage inside my body came to my awareness. Tears flowed from my eyes unbidden. Marie wiped them off and stroked my face lovingly.

She worked on my nose, finding and removing an old strain, which finally unblocked my energy. She said that tonight the damage had been structural, at the level of the physical body, whereas the last time I had a treatment, the damage had been at the emotional level.

During the treatment, I visualized the presence of Lord Jesus Christ encircling her, me, Soli, and Vivien, who was watching while holding Marie's black cat purring contentedly on her lap. I envisioned angel hands guiding Marie's hands and opening up my body to healing. I went through each part of my body relaxing and opening up to Divine Love.

The restoration of well-being afterwards felt like a healing miracle. The headache disappeared, and I felt happy to be in a physical body once again. Marie said she also felt better after working on me, adding that some people tire her, but she always feels better after working on me.

We talked for a long time about the power of this treatment, and Marie gave us the history of its development and of how she got involved in it. Marie said I needed to get some rest this weekend, and I willingly agreed. My body is *hungry* for rest!

I felt very peaceful over the weekend after Marie worked on me, and following the suggestion of the dream of November 6, I took some clay and made a goddess figure kneeling, cradling a one-inch crystal ball in her belly, symbolic of the new life incubating within me, the feminine side of God calling me to wholeness.

Akasha
Terra-cotta sculpture
Barbara Stone, 1991

Photo by Elizabeth Lozada

November 10, 1991, Meditation:

I feel Lord Jesus Christ scale himself/herself down to a body just slightly larger than mine and envelop me so that I sit fully centered within the body of Christ. I feel love pouring into every cell of my body. My arms and legs tingle.

I feel the circle of love and prayers around me.

The following dream showed a new psychic space opening up and vividly portrayed the inner attitudes that still needed work:

November 11, 1991, Dream:

I move into a new apartment, which I love very much. From the moment my friend Sara first shows it to us, I fall in love with the lavender colors and the fuchsia rug going part way up the wall.

A classmate at Pacifica and her beautiful partner, a nice man with gray hair, live a couple apartments down. I notice a sense of her womanhood being complete as she is around her partner.

I help the woman next door baby-sit. She has two children, one of which is a tiny baby. The baby goes into an embryonic state when sleeping, and the woman just lays her around anywhere. Then when she wakes up, trouble comes. Sometimes she leaves her in the toilet while washing her off, and her husband gets upset. In one scene the two children are adrift in a sailboat with a shark in the water. Soli and I rescue them.

In the ending scene, the baby makes herself into a capsule about an inch around *(like the ball in the belly of the goddess)* enclosed in tinfoil like a cold capsule. I have her and accidentally drop her. Aghast, the mother and I both look for her on the floor. She tells me I have stepped on her, and sure enough, the capsule is slightly plastered against my right foot. I take her out and look within and see that embryonic life continues. I put her up on the shelf and hope she will regain consciousness.

Later I look and see that the cells have disorganized, and she is lost. The woman does not seem to mind that she has only one child now.

Both of the children are precocious talkers. At one point the tiny one, which had been squashed, says something confrontive to me that surprises me and takes me aback, like, "Get yourself together!"

I rush to get ready for the bus and do one more thing. Then the bus is gone, and I don't know how to get to work on foot or by public transportation. In fact, I'm not exactly sure where I work.

Soli has furnished our porch with lovely furniture. I ask where it came from, and he says from my grandfather who was a healer. I am delighted.

The dream speaks on two levels: one message is that when I focus too much energy on helping others (baby-sitting for my neighbor) instead of attending to my own needs, disaster ensues. However, I am also the neighbor who is careless with her babies, her inner children. Because the baby gets stepped on, the cells in the tiny embryo disorganize, an apt metaphor for the way a cancer starts, since normal cells first get progressively more and more disorganized in their metabolic function. In the dream, this disorganized embryo confronts me with the same challenge the cancer mandated in my life: "Get yourself together!"

The neighbor woman in the dream shows remarkable detachment from the loss of her child, not seeming to mind the short time when she has only one child left. Why would such a cold-hearted image come in my dream?

One of the forces driving me has been that I have a broad range of interests and want to pursue my music, pottery, jewelry making, psychology, healing enterprises, friendships, travel, and fun, all in the same breath. Perhaps the dream suggests that since I have only one of me, not all of these potentials can be developed to their fullest, at least not all at the same time, and losing sight of one of them temporarily would not be a big problem; it would resurface later, just like the baby in the dream was talking again after having been lost.

The next paragraph in the dream reinforces the point by showing that rushing around (from always doing one more thing) not only makes me miss the bus, so I cannot get to work, but leaves me not knowing even where my work is.

The dream began with feminine images of the new internal space opening up and ended with an image of the inner male furnishing this internal space with healing.

November 11, 1991, Journal:

I absolutely rejoice that the radiation treatments to the full breast are finished! I got diarrhea on the way to the hospital this morning; my body was anticipating the effect and started it before the treatment. My weight keeps slipping down to 118 lbs. I got up to 121 last week when Soli took us out for a scallop dinner. That was the first time I have ever been happy to see my weight above my normal of 120 lbs. I have enjoyed eating heartily, but the nutrition does not stay in my system. Now those side effects will be over, and my system can normalize and return to vitality.

Today at a training session put on by the clinic where I work, a doctor put two needles into my ear during the acupuncture demonstration. I felt as if the needles in the mind-calmness and lung points straightened the top and lower halves of my head, and I felt serene. I look forward to getting acupuncture tune-ups to ease the chemotherapy side effects.

I see a new opening: we get strong at the broken places. Trouble in my life just lets in more light.

The following dream used the healing tool of humor to help me see where I needed further psychological work:

November 15, 1991, Dream:

I am with my new husband Larry in our new home. A woman enters without waiting for us to answer the door. I tell her she should wait to be asked in—we might be doing something.

I accidentally address my new husband with my ex-husband's name, then make a joke of it saying, "Now, who are you? Let me think. Pete?" I hide my eyes behind my hands and peek out. "Julie?" He breaks up, roaring with laughter.

I talk with two African Americans. One tells me of "something" that happened to a cousin twice. She was out playing in the barn, and her cousins persuaded her to "give in." I struggle through my inner defenses against looking at the issue of rape.

In the dream I am with a new inner male, Larry. I confuse him with my ex-husband, my former masculine values, and then use humor to get out of the difficult situation my Freudian slip created. Part of the confusion in my inner male stemmed from being molested when I was about nine years old while playing in the barn, the site of the rape in the dream. This buried wound needed to be uncovered and cleansed, and the dream brings this area into consciousness to start softening up my defenses against looking at the painful issue of rape.

By four days after the last radiation treatment to my whole breast, I felt great and did an aerobic workout for the first time since I had discovered cancer. I loved the feeling of my body getting hot and sweaty; I felt good, and my digestion was closer to normal. The contours of my breasts almost matched, but the colors of the breasts were reversed: the nipple on the left side was light because the radiation had bleached out the dark pigment, and the rest of the breast had a deep tan. My mood soared as my body had more time off of radiation:

November 18, 1991, Journal:

Joy floods my soul. Rainbows dance on my computer and on the wall as peace and happiness fill my being. I love myself; I love my life; I love my sexuality; I love my eyes; I love this new birth within me.

The Marcel Vogel[1] tape I listened to last night deeply inspired me, especially the part where he talked about the power of love being the force that runs the intelligence of every cell of the body. He said a spiritually developed person

---

[1]The late Marcel Vogel was a senior IBM scientist who was a prolific inventor and made many contributions to modern technology.

can make all the nutrients the body needs from just air, water, and love, with no weight loss. Could that be true?

Perhaps my struggle to get all the vitamins my body needs has been barking up the wrong tree, because I have been trying to buy and take the correct supplements to make my body work right. Instead, when happiness and love permeate every cell of my body, then maybe the cells themselves will produce what they need. All I need to do is to avoid obstructing the process by dumping in refined foods or food additives, which would upset the natural balance of my body's functioning.

I truly, truly am happy, for the first time in my life. A twinge of sadness crosses my heart for all the time I spent in darkness, loving others but not loving my own body and my own life. I rejoice that God has burned away my old life.

When I resumed radiation treatments to just the local area where the tumor was, my physical and emotional energy level fell again, but the effect was only about a third as strong as it had been for the treatments to the entire breast since the level of radiation used and the area treated were both smaller.

During this last round of treatments, my friends Harold and Starr got together with me for a healing session. When the three of us held hands in a circle to work on the radiation issue, tears instantly burst forth from me, and the thought, "It's so damn hard!" kept running through my mind.

As I cried, they said, "That's a good release." I tried to tell Harold that the poison put into my system by the radiation warped my physical body and my emotional flow. He did not accept that framework and said it was all in my mindset. He said I just had to raise my vibrational level above the effects of the radiation.

He encouraged me to watch whether people who interacted with me were relating to my weakness, saying "Oh, you poor thing!" so I could enjoy all the attention and sympathy I got from the illness, or whether they related to my strength.

Starr encouraged me to stroke myself and say, "Oh, Barbara, you're so strong and healthy!"

As we did some prayers releasing all effects of the radiation treatments from my body, I began to feel much better. We did prayers releasing all fear of the future and all fear of chemotherapy and accepted Divine Love and healing into every cell of our bodies.

I felt radiant after the session, physically and emotionally healed, and the yeast infection I had been suffering from disappeared.

The following dream that came shortly after this session showed that the shift in attitude towards the medical treatments to cure the cancer had worked its way into my unconscious mind:

November 27, 1991, Dream:

I see myself wearing a sign that says "MAXIMIZE," referring to the toxic effects in my body of the cancer treatments. Then I take that sign off and decide to wear the one that says "MINIMIZE."

I had been resisting confronting "The Black Inky Thing" mentioned in the attunement, dreading the negative energy of that package, but as the radiation treatments drew to a close, I knew I had to clean that energy out before I started chemotherapy. I finally took my mentor's advice and confronted the issues underlying the cancer in active imagination[2] in a therapy session with Edith Sullwold, a very wise woman, a seasoned and trusted guide who had been through breast cancer herself. The following images came:

Therapy Session with Edith Sullwold
Active Imagination
November 25, 1991:

I go down into the basement to find the cancer blob package, but I feel afraid to go to the corner where it is located because of the negative radiation coming out from it.

---

[2]Active imagination is a therapeutic technique developed by Jung. It uses focused attention to visualize a troubling situation and then lets that situation evolve in the mind's eye. As the process can have very powerful results, doing it under the guidance of an analyst or therapist provides a sense of safety and containment.

Edith suggests I use protection. I shield myself in light from Kwan Yin. Edith suggests I have her go with me, and I visualize her holding my hand.

I open up the package and see figures like toy soldiers on a battlefield, wounded and dying. Some are my children, my loved ones. The fear that presents itself is not as much of my own death as of losing loved ones.

I then see the form of a hooded black figure, like a Klu Klux Klan person, but robed in black. The image is Death and declares,

"I take whom I will.
You cannot control me."

As I watch, the figure turns into a spiral with the form of a fetus made of light in the center.

Kwan Yin affirms,

"Death is rebirth."

I see the image of a giant lotus flower, and each lifetime is one petal. Kwan Yin laughs as she tells me, "You can't make a lotus flower out of one petal!"

I see an image of myself dying from being burned at the stake, a direct memory of one of my deaths, and tears come to my eyes. I was burned for my spirituality. I feel afraid to let my Light shine now, because I fear the same thing might happen again.

Kwan Yin says the collective was not ready yet in that life for the teaching of Love I brought, but now the world is more evolved. She says I do not need to be afraid now.

The flames become the triple arched top of a stained glass cathedral window, and Kwan Yin says the suffering of the lifetime where I got burned was to let in more light.

Next I see the image of a large dark spiral, which turns out to be a seashell. I see myself walking up the spiral of the shell, and as I come up, the light I carry with me lights up the outline of the shell. As I emerge from the spiral, I feel wings coming out of my back at the shoulder blades.

I feel my wings expand into the corners of the room, and Kwan Yin explains I have the wings so I can fly up to God when I need to and then bring the light back down with me.

Edith instructs me to let Kwan Yin embrace me, and I do. I thank her and feel myself inside her presence, breathing me. Then I see Kwan Yin and me back in the basement. I look, and the package has turned into a golden cobra snake who tells us,

"Death is regeneration."

The visualization filled me with exhilaration from transforming my fear of death into the joy of finding my wings.

The last radiation treatment was finished the day before Thanksgiving, and I truly gave thanks for the healing miracle manifesting itself in my life through the process of the initiation.

C H A P T E R 13

Starting Chemotherapy

A dream that came during radiation therapy reflected the dread I felt about chemotherapy approaching; yet this same dream went on to link the topic of chemotherapy with a vivid image of nourishment:

November 9, 1991, Dream:

Another young woman and I are captured by some men who might harm us. They have us cornered in a restaurant. I tell them that if they kill me, at least they do me the favor of sparing me going through the agony of chemotherapy.

I distract their attention and signal to the other woman to escape while she can. As she passes me, she whispers into my ear, "Some people do most of their chemotherapy in the hospital."

I escape also and run like crazy. A man tries to follow me, but I run into the shadows and elude him. I hide in a barn, and in the darkness I smell animals. I come up close to a cow, and she motions to me with her hind foot; at first I do not catch on, but then she lays down to offer me her milk. I drink directly from her teats. I briefly wonder whether the milk will upset my digestive tract, but I know I need sustenance while hiding out. She has a larger teat with a slightly acid taste. I realize it is a penis and feel the marvel of its energy as I suck on it too.

In the dream I am trying to flee from harm, and in my waking life, I carried a fear that the side effects of chemotherapy might do harm to my physical body. The oncologist had given me a pamphlet entitled "Chemotherapy and You," which alarmed me with a huge list of possible side effects of the anti-cancer drugs Cytoxan, methotrexate, and fluorouracil (CMF) the doctor planned to use.[1]

The oncologist had assured me that the chemotherapy would be given intravenously in his clinic, not at the hospital; however, I had known some oncology patients who got so sick from their treatments that they had to do the treatments in the hospital, and the woman in the dream reflects this fear I carried.

In a twist of events in the strange second half of the dream, the person with a disease in her own breast was being nurtured directly from the breast of a cow. The curious ending of the dream, in which the cow's teat became a penis, showed creative, phallic energy flowing into me while escaping from harm. In order to deal with the unwanted side effects of chemotherapy, I had to take in a lot of physical nourishment and find creative ways of healing my body.

My new friend Starr had walked into my life during the cancer treatment, like an angel sent directly from heaven. She hooked me up with a healing group, which met regularly to pray and meditate together. Starr said that joining hands in a circle connected our energy systems like batteries in a series, and when we focused on love, we

---

[1]"Chemotherapy and You: A Guide to Self-Help During Treatment," U.S. Department of Health and Human Services, excerpts:

Cyclophosphamide (Cytoxan):
Blood in urine; painful urination; dizziness, confusion, or agitation; fever, chills, or sore throat; missed menstrual periods; tiredness; cough, side or stomach pain; joint pain; shortness of breath; swelling of feet or lower legs; unusual bleeding or bruising; unusually fast heartbeat; black, tarry stools; sores in mouth and on lips; unusually frequent urination; unusual thirst; yellow eyes and skin; redness, swelling, or pain at the place of injection; darkening of skin and fingernails; loss of appetite; loss of hair; nausea or vomiting.

Methotrexate:
Black, tarry stools; bloody vomit; diarrhea; sores in mouth and on lips; stomach pain; fever, chills, or sore throat; unusual bleeding or bruising; blood in urine; blurred vision; confusion; convulsions or seizures; cough; dark urine; dizziness; drowsiness; headache; joint pain; shortness of breath; swelling of feet or lower legs; unusual tiredness or weakness; yellowing of eyes and skin; loss of appetite; nausea or vomiting.

Fluorouracil:
Diarrhea; fever, chills or sore throat; heartburn; sores in mouth and on lips; black, tarry stools; nausea and vomiting (severe); stomach cramps; unusual bleeding or bruising; chest pain; cough; difficulty with balance; shortness of breath; loss of appetite; loss of hair; skin rash and itching; weakness.

could multiply the amount of energy available for each person in the circle to use for healing. During these group meditation sessions, I would visualize drawing Divine Love and Healing into every cell of my body on the inbreath and would visualize releasing all fear and illness on the outbreath.

During the meditation group two weeks before starting chemotherapy, my friends Gizelle and Harold instructed me to visualize the chemotherapy drugs as Divine Love entering my body, going only to the diseased parts of me, and if no disease were found, to pass on through, to eliminate the usual side effects. Gizelle also advised me to drink tons of water, like being in a fishbowl, and to eat light foods, lots of fruits and not too much heavy protein.

My dreams continued to bring up feelings I needed to work on, and a dream just after radiation had been completed brought to my awareness my guilt about leaving the stressful job at the clinic where I worked:

December 2, 1991, Dream:

I go on a home visit to see a client of mine who also had cancer, entering through her back door, and she is not home. Others are there, including a Pacifica schoolmate and relatives of my other clients. I leave.

Later I talk to my client about the importance of keeping our visits. She looks me straight in the eye and asks me if I am leaving the clinic. I say no (though I know I am lying), but that I am cutting down to two days a week.

She looks anxious and says sick people like herself need the chance to talk with me to relieve themselves.

As the dream said, I did cut down my work at the clinic to two days a week when I began chemotherapy, a compromise between my need to be in a less stressful situation and my loyalty to the needs of my clients; but I still felt guilty about my plan to eventually leave entirely.

My energy had bounced back again by a week after completing radiation, and I felt terrific. I flew to California two days before my regular session of doctoral classes and met my older sister in Santa Barbara on Saint Barbara's Day, December 4, which was also my birthday. We spent a delightful day together and talked nonstop, comparing notes on all we had learned about holistic healing and life in general since we last saw each other.

My sister, a professional massage therapist, worked on my body for a long time, gently coaxing out the tension that remained in my muscles from all I had been through. She told me, "I got this shirt with beautiful teal and orange colors to wear while I worked on people, and then all of a sudden I saw that the design on it was *a huge snake!* Yikes!" When I told her about my snake dreams and the healing properties of the snake image, she gave me the shirt for a birthday present.

At school that weekend, I was shocked to find out that a student in one of the other classes had committed suicide during the previous month.

December 6, 1991, Journal:

How could a fellow student do that (suicide)? I know the feeling of longing for death, the feeling of wanting the sorrow in life to end, but I feel resolution of my ambivalence about being in a body.

As I face death squarely, I also let go of my desire for a way out of my suffering. In fact, I release my suffering; I let it go. I'm tired of dragging it around.

I feel more alive since the powerful and transformative active imagination with Edith. I do not fear death or the mode of my death. I live more abundantly now than I ever have, and I feel better right now than I ever have before. Amazing! I feel Light radiating out from me.

December 7, 1991, Journal:

I feel total peace and relaxation in my body. As I curled up in the fetal position, I saw an image of the hands of God holding me like a womb, protecting me. I feel Perfect Comfort inside. Joy radiates from my loved and loving heart.

Dreams on two consecutive nights just before chemotherapy once again brought up the topic of sexual abuse and linked it to snake imagery:

December 9, 1991, Dream:

Grandma Bontrager says she read a book once about sexual abuse but did not like it. Grandpa said somebody went out in the woods once looking for a snake, but they never wanted anything to do with snakes.

December 10, 1991, Dream:

At Grandpa and Grandma Bontrager's old farm, I look out the front door and remember all the good times of playing there.

A doctor is giving vaccinations with the bite of a spider and two snakes. I get in line, then think I don't want the spider and snake to bite me. Then I see he does it with a needle dipped in the venom: a stick for the spider and an IV for the snakes. He says I'm not ready yet, so I don't get it.

I go to a banquet in my honor. They serve "Fillet of Sole." I take some but then do not eat it because I go back to my workplace to get my boss, who decides not to come. The psychiatrist says to get myself over to the banquet. I am wearing a green dress and my hand-woven jacket. The banquet is also about my change of jobs.

The association of chemotherapy with the venom of the snake bite in the second dream is obvious. Since no evidence of any other cancers had been found in my body, the adjuvant chemotherapy recommended for me was indeed like a vaccination against future occurrence of cancer. In the dream, I felt aversion to the venom involved in the healing process, and the inner doctor reflected my feeling that I was not yet ready to be vaccinated IV with the snake venom, chemotherapeutic agents.

In the final paragraph of this dream, I miss out on partaking of the nourishment at the banquet in my honor because I go back to get my boss at my workplace. I was having difficulty accepting the reality that to fully nourish myself, I would need to leave the stressful situation at the clinic; however, the boss (representing the clinic job) could not come to this banquet; my internal psychiatrist urged me to get back over to the banquet place of honor and nourishment, a place of growth (the color green) and beauty (my handwoven jacket).

To get my body prepared for the onslaught of snake venom, I scheduled my first acupuncture treatment the day before the first chemotherapy.[2] I really liked Suki, the acupuncturist, who was a cancer survivor herself. She had undergone six months of radiation as a teenager when she had Hodgkin's disease and had tattoos on her chest too, and we became good friends during my weekly treatments throughout chemotherapy. I felt relaxed and peaceful on her acupuncture table and could feel the tiny needles, left in twenty minutes on the front and then twenty minutes on the back, stimulating my body's energy meridians.

The banquet in my honor in my dream of two nights earlier had been about my change of jobs; ten months later, I accepted Suki's invitation to join her association of holistic health practitioners, opening up a private practice in her office suite. When I first met her, I "never dreamed" we might eventually work together, but my dreams had clearly outlined the necessity of changing jobs and the nourishment that would come from such a shift.

The work I have done at the acupuncture association has deeply satisfied my desire to be in a setting where I could incorporate all my healing tools, truly a nourishing "Fillet of Sole," fish cooked for soul pleasure.

The evening before the first chemotherapy session, I took the first tablet of the anti-anxiety medication Ativan, prescribed by the oncologist, supposedly 1 mg of serenity. I had been very curious to find out how a minor tranquilizer would feel in my body and was surprised that I felt no immediate effect. I had already felt peaceful ever since the acupuncture treatment, and as far as I could tell, taking the Ativan tablet did not improve my state of mind.

December 12, 1991, Journal:

Tomorrow I start chemotherapy. I have been dreading and anticipating this day for not just months, but years. The thought of chemo has always terrified me, and now I am facing that fear head on.

---

[2]For a brief description of acupuncture, see Appendix D.

> Actually, I'm not afraid. I feel the presence of my guardian spirits and angelic friends all around me and feel the love and prayers of all the people in my life who care about me. I am grateful for the opportunity to rid my body of any cancer cells that might be left. I want to stay alive for a long time yet if possible. Nevertheless, if today where my last day of life, I would be pleased with the way I have spent it.
>
> I have a feeling that I will get through the chemotherapy pretty well. My soul is so happy inside, I could cry for joy. Truly all is well in my world!

The next morning, on Friday the 13th of December as I quieted my mind for meditation before heading off to the oncology clinic, I felt the presence of God was saying to me,

> Today is a day of purification for you. While the doctors are putting the medicines into you, simultaneously, angels will be around you infusing you with light, purifying your spirit.

During the actual treatment, which took about an hour to get all of the saline solution containing 950 mg of Cytoxan, 65 mg of methotrexate, and 950 mg of fluorouracil through the IV in the top of my right hand, I was grateful for Soli's healing presence beside me, holding my left hand. I gave thanks for this cleansing treatment to purify my system of cancer, imaging the yellow chemotherapy drugs as Divine Love, and visualizing being surrounded by angelic presences.

The Compazine (a major tranquilizer) and Ativan I had taken by mouth before the treatment made me drowsy, and I dozed off on the ride home and then fell asleep on my living room couch.

Waves of nausea soon began, and I kept throwing up every couple of hours. Soli gave me a full body massage, and Marie arrived mid-afternoon to work on me. Marie said my energy was very lopsided, all shifted toward the right side of my body (where the IV had gone in), and she worked to rebalance my energy system. I had to interrupt the craniosacral treatment to vomit, and while I was dry heaving in the bathroom, I heard Soli crying with Marie in the kitchen.

The nausea was much worse than I had expected. When I had first told my mother that I was going to undergo chemotherapy, she had suggested I get some marijuana to help with the nausea, but I ignored her advice. I was surprised that my own mother would suggest I break the law, and besides, I had no idea where to get any. I knew my clients had access to drugs, but I could not possibly put them at risk by asking them to locate some for me. Although I saw drug dealers all around the clinic where I worked, I could not put my agency at risk by trying to buy any pot there either. I could just imagine the newspaper article that would result:

SOCIAL WORKER ARRESTED ON DRUG CHARGES

(AP) Barbara Stone, LICSW, 43, was arrested today on charges of possession of illegal drugs right in front of the clinic where she had been employed as a psychotherapist.

Undercover narcotics agents spotted her leaving the agency (which fired her following her arrest) and trying to buy marijuana from a car mechanic, a housing maintenance worker, and a grandmother before she finally located a real drug dealer.

Stone' s only comment to newspaper reporters present at the scene of the crime was, "It's all my mother's fault!"

My chemo started in 1991, and my home state of Massachusetts did not become the thirty-fifth state to enact a law to make marijuana available for medical use until 1992 (Grinspoon & Bakalar, 1993). I had never mentioned the possibility of getting any cannabis in leaf form as marijuana or in its chemical form as THC tablets to my oncologist, and I had never even talked with others about using cannabis and had not read up on it to find out that for maximum effectiveness to relieve nausea, I would have needed to use cannabis *before* the chemo (Grinspoon & Bakalar), like the anti-nausea drug the oncologist prescribed to use the night before each treatment.

However, the intensity of retching over the bathroom toilet soon changed my mind about engaging in criminal activity, and I decided I needed more help with the nausea than my prescription drugs were giving me. I wanted some cannabis. I decided that if a police officer tried to arrest me, I would throw up on him!

Soli tried calling different friends we knew to find some marijuana, but nobody had any. Vivien and one of her friends went out searching for some, but they could not find any either. We were all totally inexperienced at locating dope! Soli ran to the corner store and bought a corncob pipe so I could smoke the pot if we ever found any, and meanwhile, I kept throwing up. I felt as if I had a case of the 24-hour flu, and in between episodes of vomiting, I slept most of the time for the first thirty hours following the treatment. I had asked the doctor for the special anti-nausea medication Zofran (ondansetron) for my first treatment, but he did not give it to me because he wanted to wait to see whether I really needed it or not.

The following evening some true friends came through in a pinch by bringing me a joint of marijuana they dug up from a friend of theirs. I took one long drag on it, and my stomach settled down. As we talked with our friends, I gradually woke up and joined hands with them in a group prayer for my healing. As we meditated together, I gradually began to feel better. I was surprised to find my body working again.

By the second day following chemo, I realized that the Ativan prescribed before and after the treatment had scrambled my brain, and I had no sequential memory of the events of the previous two days. I could not think clearly and had to keep asking Soli what had happened. Later I read that Ativan is given to induce amnesia (Markman, Theriault, & Williams, 1991), on the theory that if the oncology patient does not remember the episode of nausea, the body will be less likely to develop a conditioned response of nausea to the same drug when it is given the next time. I suddenly realized that my clients on Ativan were the ones who had the most trouble remembering their appointments with me, and I felt sympathy for the mental confusion this drug produced.

The night of December 14 I dreamed that I was trying to print out my paper, but other people's papers kept coming out instead, a reinforcement of the theme of needing to get more focused on my own needs than on the needs of others, like the dream of having only other people's clothes in my closet.

Marie came over the next night to do another craniosacral adjustment on me. During the treatment, which was done in silence, images came to our minds. When we shared notes after the treatment, we found that we had spontaneously visualized similar things. She had seen an eagle head in me, and at the same time, I had seen in my mind's eye the image of a ray pointing down, which then fanned out in an arc with an opening at the top, like spread eagle wings.

As she worked on my sternum area, I felt the sensation of liquid flowing inside of my bones, and Marie felt the energy move too. At one point during the treatment, I felt intense ecstasy, and at another point, tears flowed as I saw the heart of the goddess full of sorrow, holding the sorrows of the world. After the treatment, I felt lighter and energized, and from then on, I was up and around again, with a normal energy level. I even managed to finish all three of the papers due for my doctoral program and turn them in before the deadline for the winter term.

Soli, Vivien, and I headed 900 miles west the following week to spend Christmas with our parents, and while visiting there, I had the following dream:

December 25, 1991, Dream:

The former pastor of our church is pacing outside the sanctuary during Sunday School. I ask if he knows of the illness I am going through, then remember telling him already.

He has symptoms himself and wonders whether or not he has cancer. I run through the list of cancer symptoms, but he has none: his symptom is that he is *horny*. I say this symptom does not indicate cancer; instead, it shows his vital life energy is at work. I tell him to visualize what form of life energy is calling his creativity.

The dream is correct that feeling lots of sexual energy is not usually a sign of having cancer. On the contrary, researchers such as Carl and Stephanie Simonton (1978) and Lawrence LeShan (1977) have linked the presence of cancer with symptoms of depression, which tends to lower all forms of life energy, including interest in sex. But in this dream an inner spiritual male figure is experiencing life energy calling his creativity, just as the process of going through cancer treatment drew my spirit into new creative healing ventures.

When I opened up my Christmas gifts, I was absolutely delighted to find that my family had showered me with lots of purple clothes made from 100% natural fibers, a welcome addition to my wardrobe. I wanted to saturate myself with amethyst jewelry and purple clothing to draw more energy into the process of spiritual healing that was going on inside my soul.

Just after Christmas my mother took me to see a doctor who had used nutritional supplements to help her with several health problems. This doctor checked me over and said the only two things he found wrong with me were slight hypothyroidism and too much yeast in my body. He thought that the most common cause of breast cancer was yeast/fungus/mold in the body and encouraged me to avoid milk, ice cream, and sugars. He said anything fermented was all right to eat, like cheese, yogurt, rice milk, and soy milk.

I could not imagine how yeast could make cancer, but I hated the vaginal yeast infections I had been prone to for many years and definitely wanted to get rid of them. During the previous year, I had suffered three severe yeast infections plus a new problem, a recurrent fungal infection on my thumb.

To help get rid of the yeast, the doctor prescribed a nutritional supplement called Zymex (containing defatted wheat germ, lactose, tillandsia, and beet root) for an entire year. He also recommended taking tablets of organic iodine for the rest of my life to give me energy since my thyroid was sluggish. I decided that even if the supplements did not help me, the only thing they could hurt would be my budget, so I bought a six-week supply to see how I felt after taking them.

The vaginal yeast infection problem receded immediately with the supplements and diet changes, and I felt that the iodine tablets helped to perk up my metabolism, giving me more physical energy. The fungal infection on my thumb disappeared and never came back. I decided to take the Zymex tablets for the whole year, wanting to do all I could to improve my physical health.[3]

On the trip back home to Massachusetts, I got a runny nose from getting in and out of the car so much in the cold winter weather. Soli

---

[3]A book which came out two years later, titled *Beating Cancer with Nutrition* (Quillin, 1994), pointed out several principles that support this doctor's theory. Since research has not found a "magic bullet" cure for cancer, a teamwork approach is needed. Aggressive medical interventions like surgery, radiation, and chemotherapy can be combined with diet changes to promote conditions in the body that will selectively starve the tumor, nourish the healthy part of the body, and bolster the immune system. Since cancer tumors thrive on sugar, eliminating sugars from the diet makes sense. Anecdotal reports also link fungus to cancer. Yeast, a relative of fungus, is also involved in cancer mortality. Infection is one of the main causes of death in cancer, and half of these infections come from yeast overgrowth, mostly Candida. Iodine deficiency is also implicated in cancer: "low thyroid output substantially elevates the risk for cancer" (Quillin, p. 116). Quillin recommends that cancer patients eat kelp or other sea vegetables to help supply the diet with iodine to improve thyroid function.

asked if I wore a hat when I got out, and I said no. He asked whether I had a hat I liked, and I told him that they all mess up my hair. He responded, "Well, Barbara, it's a matter of priorities. If you get sick and die, you won't have to worry about how your hair looks!" The truth in that statement caught my funny bone, and we giggled together for a long time.

On New Year's Eve I resolved that in the coming year I wanted to have more fun, more joy in my life, more humor, and more enjoyment of each day as it came.

A vivid image of cancer treatment came in a dream on New Year's Day:

January 1, 1992, Dream:

Back in the woods, at a renewal center, I am swinging from the trees by ropes, a wonderful feeling of soaring. I can control my landing spot by the angle of my "sky dive." I see people and animals below.

A bear grabs the end of my rope and will not let go. I shake the rope, but he hangs on. I swing it so it thumps him on the head, but the bear still persists. I thump more times, bashing his head, but still he hangs on. I hit him on the tree so many times that piece after piece of his body falls off, finally leaving nothing left gripping the rope except his jaw. I feel bad that the bear was destroyed in the process, but I could find no other way to avoid the bear hurting me.

When I finally want to leave, I cannot find my bicycle. As I look in the hill slopes below, a young boy walks with me part way. He is holding a pet snake, which I see bites him. I ask him whether he was bitten by the snake, which I am pretty sure is poisonous because of its brown color and stripe pattern, and he says no; however, in the process of answering, he breaks the head off the snake. A green beneficial snake is following him.

Cancer is a bear. It is powerful, tenacious, unpredictable, and surly. When I discovered the tumor in my breast, the bear hanging onto the end of my rope, I thumped it on the head with surgery and radiation, but it still hung on. Each chemotherapy session banged the body of the bear against the tree one more time, destroying the cancer piece by piece until nothing was left but the jaw bone of the bear. The method was violent, but the cancer had to be destroyed to protect me from harm. The fact that I noticed the bear while I was soaring in the air at the renewal center points to cancer being a part of my initiation, my renewal.

After banging the bear, I could not find my bicycle, a means of transportation one powers by balancing oneself on top; and I could not recover from the cancer treatment solely under my own steam. I needed to let in lots of love and support from other people.

In the ending of the dream, my young inner male is bitten by the pet snake he holds, and my ego thinks the serpent is poisonous; however, the inner male breaks the head off of this creature, breaking the back of the destructive aspect of snake symbolism. The process of healing through surgery-radiation-chemotherapy, coined the "slash, burn, and poison" method by Dr. Susan Love (1993, p. 50), is crude and feels dangerous and poisonous at times. However, just as the body of the bear was broken off its head, the dream showed the head of the poisonous aspect of the snake being broken off; then, a beneficial green snake followed, the healing and renewal aspect of snakehood.

The second chemotherapy session came on January 3 and was more difficult than the first treatment:

January 3, 1992, Journal:

My sleep was restless last night in anticipation of another chemotherapy session today. I dreamed of being late for a doctor's appointment. Then Soli and I both forgot to set the alarm and overslept, giving me no time to do yoga and meditate this morning, only time to hop into the shower, dress for work, and eat breakfast.

I had my blood drawn before work to check my cell counts. I hate getting my blood drawn, partly because the needle itself hurts and partly because it feels so inhuman to take people's blood out of them. We need it! The needle spot in the crease of my elbow hurt all morning and was bruised when I took off the bandage.

> I noticed myself becoming shaky this morning as I thought about calling the psychiatric nurse to see whether my blood counts were high enough to have another chemo treatment this afternoon. I especially noticed my anxiety when I signed my name and spelled "Barbara" wrong. However, I was delighted that my white blood cell count was 4,000, actually in the normal range. The acupuncture is working! The platelet count was also good, 202,000.
>
> During the chemo session at 3:30 today, the doctor first put the anti-nausea medication Zofran into the IV since I had so much vomiting from the first treatment. I started to feel dizzy. The nurse said it should not make me feel that way, but I got lightheaded, and my vision began to cloud and go black. I told her I felt I might black out, and she had me put my head down between my legs. I was scared of the feeling inside at first, which was a bit like anesthesia.

The only good part about the session was that a dear friend of mine got her chemo simultaneously, and we got to talk over all our symptoms with each other and with the nurse during the ninety minutes while the IVs were going in. Having the mutual support of my friend during the bear-banging process helped me feel less isolated by the disease.

That evening I felt sleepy and dozed on the couch all evening, listening to healing tapes, but I did not feel sick. I did feel cold, though, and had trouble getting warm enough, even right by the wood burning stove with a heavy feather comforter on me. My internal thermostat seemed to be on the blink.

I was pleased with the initial action of the new medicine Zofran because I had only a slight tinge of nausea, which lasted just a couple of seconds.

Soli was shocked that the Zofran cost my clinic $202.00, astronomically more than the $3.28 for the Ativan. However, even though I never vomited with the second treatment, nausea set in on the second day and lingered for four more days:

January 6, 1992, Journal:

Nausea. The pit of my stomach aches and wants no input. I know I should drink ginger tea, but the thought of drinking *anything* nauseates me further. Tears well up unbidden. I can't face the thought of going through four more of these chemotherapy "treatments."

Here I am, a primarily Pitta body type who is extremely sensitive to any impurity in my system, and they put something called Cytoxan into me. This time, the nurse told me that the CMF medication by-products I will excrete for the first 48 hours are so volatile that if mixed with bleach, they produce toxic fumes. She told me to clean my toilet bowl and my clothes with only soap and water. Otherwise I would turn our bathroom into a health hazard and make our septic system into an environmental hazard. I am also advised to urinate at least every four hours so the by-products do less damage to my bladder. The nurse said to use a condom if having intercourse during the first 48 hours, or else the man could get a skin rash from the toxicity in me. She must be crazy to think a person feels like having sex when her body has been assaulted by the healing cell-killer medicine!

If the cancer feels anywhere nearly as bad as I do from the onslaught of this treatment, it will definitely give up and be wiped out.

A curious dream came at the end of the five days of feeling miserable after the second chemotherapy treatment:

January 8, 1992, Dream:

Janice is three months pregnant. She tells us she just had an operation that cost several thousand dollars to fix something in her breast, so she would not be able to nurse her baby. I see the scar on her left breast.

I feel sad she will not be nursing, but she says she probably would not have been able to nurse anyway, and this operation ensured the success of the pregnancy.

When I awoke from the dream, I asked myself what within me had been fixed so I could not breastfeed, in order to ensure the success of my internal pregnancy. My general pattern of serving others while ignoring myself came to mind, having only other people's clothes in my closet and only other people's papers coming out of my printer.

If I would have poured as much of my time and energy into my clinic job as I felt it needed, that output of energy would have endangered my recovery from the cancer treatments. In the dream, the scar was on my left breast, the site of the lumpectomy, and the dream helped me get through the guilt I felt about cutting back at the clinic and eventually leaving, because I saw withdrawal was necessary to protect my recovery from cancer. The milk faucets had to be sealed because I needed to keep my energy inside to use for my own healing.

I made myself a shirt that said "NO" in letters six inches high on the outside but said a big "YES" on the inside of the shirt, because when I said "no" to others, I was really saying "yes" to myself. I let go of the guilt I had been feeling about spending so much time on my healing and dedicated myself even more deeply to my own care.

The dream that followed two nights later showed my unconscious mind had absorbed and integrated the necessary attitude:

January 10, 1992, Dream:

My husband's good friend is visiting. Because of attending to my own needs, I pay little attention to taking care of him.

The friend is restless, not sleeping well. I am in bed with Soli, and as I lay directly on top of him, I feel waves of love for Soli encompassing my entire body. I take off all my clothes to make love together.

Just then his friend comes through our room on his way out, with his bags packed, saying he needs to get back to his base. I apologize for not paying much attention to his needs during his stay, feeling some guilt over his leaving early.

He says there is no need to apologize and that he would have felt guilty if I had spent the psychic energy I need to heal my own body on him.

The attitude of dedicating myself to my own care was reinforced in the dream by my union with Soli, who as my husband was also a symbol of my inner male, my inner caretaker. The day of the dream I cared for my physical body by getting an acupuncture treatment, and I could physically feel the acupuncture strengthening my internal organs; the treatment left me feeling terrific.

As I listened to a healing visualization tape afterwards, an image came to me that the chemotherapeutic agents were a solvent to dissolve the smear of cancer superimposed over the surface of my strong and healthy body. I visualized myself, standing on a beautiful mountain in radiant health and strength, with long wavy hair blowing in the wind as I lifted my outstretched hands to the heavens, pulling Divine Love into my body and into the earth.

Meditation group met the next night, on January 11. I had read an article, which named this date the "11:11" for the first four numbers in the date. This newsletter, which was printed on purple paper, claimed that this day was a chance for awakened beings on the earth to drop the illusion that the people here on this planet are separate from each other and to remember that we are really all One. According to the author, if we were able to transcend our individual egos and come together in unity on this date, humanity would collectively be able to move into a higher octave of the evolutionary process (Amaa-Ra, 1991).

After this meditation session, I *knew* that the cancer had quietly gone from my body. I had opened myself up to healing from the universe. I did not feel any physical sensation; I just knew with a deep intuitive certainty that the disease was no longer present in my system. I felt radiance in my being, a deep, quiet energy that came out from my core, shining like the sun.

The next time I got a cranial manipulation treatment, Marie observed, "Well, one good thing is that I don't smell cancer in you anymore. I can tell when patients have cancer by an odor they carry." I asked when she did smell it, and she said it was so strong right after the surgery that it was difficult for her to tolerate, but it got gradually better with the chemotherapy, which she felt had helped me. I then told her that I myself felt the cancer leave on 11:11 (Jan. 11), confirming her smell diagnosis.

I did not feel I really needed the remaining four chemotherapy treatments, but I decided to finish them out for added protection. I wrote in my journal,

No longer do I think of myself as a person who has cancer. I *had* it, and now it is gone.

CHAPTER 14

Kirlian Photography

The three main treatment modalities I used to fight the cancer were surgery, radiation, and chemotherapy. To these healing tools, I added some secondary interventions to ease the unwanted side effects from the medical modalities and to boost the function of my immune system. I used so many tools to fight the cancer that sorting out the individual effect each one had on the whole treatment was impossible.

Which treatment cured the cancer: the surgery, the radiation, or the chemotherapy? I will never really know, but I wanted to get an idea of what each treatment was doing to the energy in my body. I also wanted to get a general idea of whether or not the holistic healing methods I tried were actually helping me. In addition to documenting my reactions through self-report in my journal, I wanted some type of measurement, which might be a bit less subjective.

While I had been studying in Zürich in 1987, I had done some extra-curricular experiments with Kirlian photography in a private studio. These pictures were taken in a darkroom using a Kirlian device, which was connected to a metal plate. A piece of unexposed photographic paper was placed on top of this metal plate, and then I put my hand on top of the photographic paper. When the technician turned on the machine, I loved watching the luminous blue sparks dancing around the edges of my hands. The technician then developed the photos in his bathtub, and I was intrigued by the resulting pictures, which showed the outlines of my hands and feet with rays of energy called coronas feathering out from the edges where my body made contact with the photographic paper.

I wanted to see how the treatments for cancer would affect Kirlian photographs of my energy.

The aspect of Kirlian photography that fascinated me the most was the "phantom leaf" phenomenon. When a portion of a healthy geranium leaf was removed and the leaf was re-photographed, the Kirlian picture sometimes showed an outline of the entire leaf. Whatever a Kirlian apparatus photographs, it is not something purely in the physical world, because it can photograph part of a leaf that is not there. It measures an energy that is not visible to the human eye, an energy based on form rather than matter. This invisible energy form seems to surround and interpenetrate the leaf, being present even if the physical form of part of the leaf is removed.

At this point, the reader may want to ask, "Just what exactly does Kirlian photography measure?" In simple words, the force measured could be called "The juice that gives us life" (Steiner, 1977, p. 13). A more technical definition comes from Nikola Tesla, whose Tesla coil photographed this energy in the late nineteenth century. Tesla defines the force Kirlian photography measures as a "bio-electrical resistance which changes according to the physio-chemical and electro-dynamic reaction in the living specimen" (Steiner, p. 17). For further information on the theory of Kirlian photography, see Appendix B.

The physical body needs a flow of energy through it to run its physical and emotional mechanisms, like the electricity that is needed for a city to operate. If this "electricity" is completely cut off because of an engine malfunction (a heart problem) or a serious break in the lines of energy (gunshot wound to a vital area), a total "blackout" can occur: death. A living body minus its "electricity" is a corpse, two very different items!

Sometimes the power will fail in only one section of the town, like kidney failure or the pancreas shutting down its production of insulin. Another possibility is that if the generator does not have sufficient power, a "brownout" may occur in one region of the body, and this pattern will show up on a Kirlian photograph as an area with spotty or weak sparks.[1]

---

[1]The analogy with electricity comes close to describing the Kirlian energy in lay terms; however, the energy is neither electricity nor electromagnetism. The closest one word to accurately describe it is "bioluminescence" (Ostrander & Schroeder, 1971, p. 211). In other words, it is the body's "glow."

Cancer is a disease in which DNA has malfunctioned and instructed a cell to reproduce much too fast. Each of the 60 trillion cells in the human body contains a thread of DNA, which is like a long, flexible staircase that contains 50,000 to 100,000 genes stored on 23 chromosomes. These DNA strands are under constant attack from the pollution and radiation in our environment, and each cell takes an average of 1,000 to 10,000 "hits" or breaks in the DNA each day. Usually these DNA errors do not turn into cancer since the body has a repair system, an enzyme called "DNA polymerase," which moves along this spiral staircase repairing the breaks.

> Geneticists have estimated that each DNA molecule contains about the same amount of information as would be typed on 500,000 pages of manuscript. Imagine if trillions of times daily, you had to type a half million pages error-free. A mistake can lead to cancer. (Quillin, 1994, p. 99)

Perhaps any area of the body that does not get an adequate supply of energy may not be able to accurately rewrite its daily DNA damage, eventually resulting in cancer forming in that area of the body.

A Kirlian photograph is considered capable of detecting an energy imbalance before this same disruption in the natural flow of energy in the body leads to physical illness (Ostrander & Schroeder, 1971).[2]

Semyon and Valentina Kirlian, the husband and wife team who developed this tool behind the iron curtain, found that Kirlian photography could be used to diagnose illness *before it showed up in physical form*. They found that if a plant had a disease, the Kirlian photographs of the plant started to show a faded, distorted pattern of energy several days or weeks before the disease could be detected in the plant itself.

---

[2]An energy imbalance may come in different forms. In one Kirlian experiment, a finger that had been recently broken showed a much brighter corona than another finger on the same person, which was not broken (see photos in Moss, T., 1979, p. 92). The body naturally sends its energy to a wounded part to try to heal that spot; however, in putting all of its eggs in one basket, other baskets are left empty. In another experiment, injecting one hand with a drug made that hand have very bright coronas; however, the other hand totally disappeared from the Kirlian photograph (Moss, T., 1979). In other words, energy is a little bit like money: if you spend it all in one place, some other part may go bankrupt. In a healthy state, energy is balanced in all parts of the body, leaving no empty baskets or overdrawn accounts.

A Kirlian photograph of my feet taken in February of 1987 had shown a few open, spotty areas in the outlines of my feet and a blotched area on the left footprint.[3]

The person who took the photograph asked me what was wrong in the area of my belly since the black globs on the print of my left

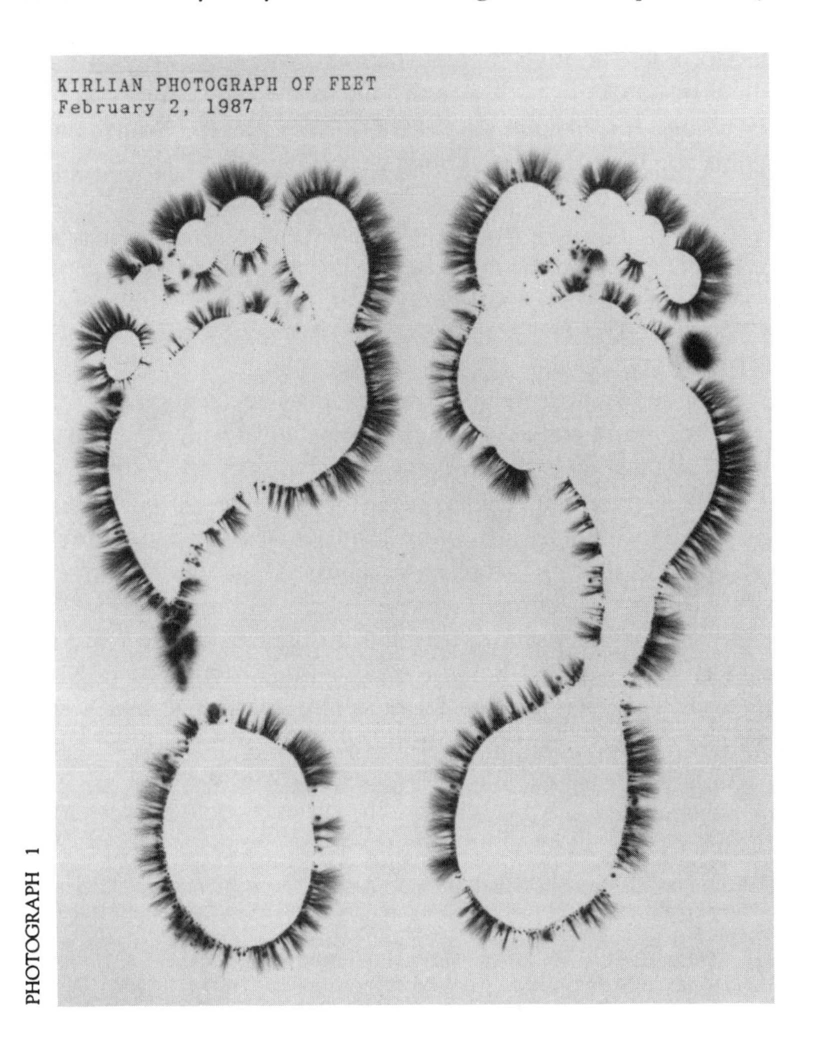

PHOTOGRAPH 1

[3]According to the Kirlian photographer, any area in the Kirlian outline of the body that is open is vulnerable to being damaged. In a healthy state, the outlines should be evenly filled in. Also, a large glob or a splotch instead of a fine ray shows an energy blockage, while tiny little dots show toxins. He also said that the feet indicate the state of the physical body, and the hands indicate the state of the mind.

foot just above the heel fell right below the area which corresponded to my waist.[4]

I did not know of anything wrong then, but when I returned to the United States later that year, I had my first brush with cancer: in September, 1987, a pelvic exam found a fibroid tumor the size of an orange in my uterus, and my Pap smear showed mild disorganization of the cells, the beginning of cervical cancer.

The progression of cancer in my cervix was promptly halted by freezing the surface of the cervix (cryosurgery) so that it would slough off all of the aberrant cells. The doctors decided to just watch the fibroid tumor since it was not dangerous by itself and since I had no plans to ever get pregnant again. They said it would likely shrink all by itself when I went into menopause since fibroids feed on estrogen.[5]

I wondered whether the black glob in Kirlian photograph number 1 had reflected some imbalance of energy, which might have contributed to the beginnings of this cervical cancer.

The thesis project I had formerly been planning included studying the effects of healing methods on the energy fields of a dozen subjects using Kirlian photography in addition to self-report. Dr. Peter Nelson, a research professor at Pacifica, helped me build my own Kirlian apparatus from a diagram I found in a book (Martin, 1974, p. 72), but it would not work from that diagram. Dr. Nelson moved to Australia before he could figure out what was wrong with the diagram, so I took it to my friend Mike, who installs security systems. He and his electronic engineer friend Frank worked on it a bit, trying to make it functional. When I found out I had cancer, I called up Mike, and he and Frank promptly put their brains together and adapted the system so it ran (see Appendix C for diagram) so I could look at the changes

---

[4]Areas shown in a Kirlian photograph of hands or feet correspond to the reflexology areas for hand and foot massage. Roughly speaking, the foot contains a nerve map of the entire body, with the brain at the top of the toes, then descending down the foot, organ by organ in order until the foot area falls at the bottom of the heel. For a detailed map of the areas, see the chart "Foot & Hand Massage Reflexes—Rainbow Color Coded" from Silvermoon Wholistic Research, P.O. Box 82542, Tampa, FL 33682, 1987 Edition.

[5]In subsequent exams, the fibroid first got bigger, growing to the size of a grapefruit by 1991. Then my hormonal balance shifted with the chemotherapy in 1991–1992, and menstruation slowed down and then stopped. When I went to my gynecologist in August of 1993, I had not had a period for the past three months, and I still had a mass in my uterus. She sent me for an ultrasound to get another look to be sure the mass was not a tumor and remained a harmless fibroid. This ultrasound showed that the fibroid had shrunk back down to its 1987 size.

in my Kirlian photographs as I went through the treatments for cancer.

The night before radiation treatment began, September 25, 1991, Frank and Mike met with Soli and me at Mike's house to do our first experiments photographing our Kirlian handprints. We only got a few sparks, which did not show up well on the photographs developed from that first night, but I felt exhuberant that at least we had gotten started.

When Soli and I got the box home, however, it would not work at all in my pottery studio, which I set up as a darkroom. Soli and I sputtered around for awhile, trying different ways to get it going. With his engineering expertise, Soli helped me think through the possible problems that might prevent the system from functioning properly.

I could tell by the feel that the ion flow generated by the system was not strong enough to generate the luminous blue sparks I had seen in the studio in Zürich, and my library research on Kirlian photography (Steiner, 1977) gave me the idea of photographing just fingertips instead of the whole hand, to concentrate the ion flow into a smaller area instead of diffusing it throughout the whole hand.

I wanted to observe both physical and emotional functions, and both of these areas are regulated by brain states. Since the fingertips correspond to the hand reflexology area for the brain, photographing just the tips of the fingers made sense.[6]

When I tried the apparatus with just the ends of my fingers touching the photographic paper, beautiful crowns of luminous blue light rays streamed out of each fingertip from the collision of the ions generated by the Kirlian device with the ions my own body was producing. The sparks reminded me of the light pattern produced by two live wires touching, and I felt absolutely thrilled to see this spectacular light pattern.

I had been waiting very impatiently for two years to see the Kirlian box finally working, and I could barely contain my excitement. The previous day I had been in the pits of despair from the depression induced by the radiation treatments, and the success of this project lifted my spirits. Soli and I together celebrated the beginning of find-

---

[6]The correlation between corona pattern and mental state has been well-established by other researchers (Iovine, 1994; Steiner, 1977; and Moss,T., 1979).

ing a new window we could use to peek at the effects all the treatments and therapies I was using would have on the glow of light coming out of my fingertips.

I did not set up the Kirlian photos as a scientific experiment because I could not possibly have controlled for all of the variables influencing my brain state at any one time. *My goal was not to prove or disprove any theory, but to get well as fast as possible.* I wanted to be able to sample my energy state at different times to monitor what the side effects of radiation and chemotherapy were doing to me and to see whether or not the extra healing methods I added seemed to be helping. I have a lively imagination, and I wanted to see whether some of the ideas I collected were just "wild hairs" or whether they really had a beneficial effect on my energy.

One of the things I tried was wearing a magnetic belt around my waist, supposedly to enhance the body's energy flow. The Kirlian pictures of before and after wearing the belt were both erratic, but I was so sure the belt was working (its pamphlet said it would) that I thought the pictures while wearing the belt looked better. However, when I showed the photos to my friend Rebecca, she preferred the ones without the belt.

I then tried taking more pictures, reversing the order of the photos with the belt on and off and varying the time lengths I wore it. Finally, one day when I had worn the belt for ten hours, photographed my fingertips with it on, and then took it off, I felt a flow of energy surge through my body and deep relaxation set in as the belt came *off.* The Kirlian print with the belt off was clearly better than the one with it on had been, so I stopped wearing the belt.

I often coupled interventions together, like meditating during an osteopathic craniosacral treatment and then doing a group meditation, meditating during acupuncture, and meditating while I was in the sauna. Distinguishing whether the improvement came from the meditation or the other treatment was impossible; I could only surmise whether or not the combination seemed to be helping.

General trends emerged from the 140 Kirlian photographs taken, and I have included representative examples from these trends in the 21 photos in this chapter. Although I give my opinion on what factors might have caused the change in sparks for each set of photos, I leave the reader to form her own ideas of what the changes might have come from.

Surprisingly, the photos I learned the most from were the ones which did *not* fit the prominent patterns, and I have also included those exceptions.

A healthy Kirlian print should show a dark inner circle called the corona surrounding the edge of the contact area, rays emanating out from this corona, and a full imprint of the fingerprint in the center. The corona should be complete, with no breaks; the rays should be balanced and even around each circle. The five fingers of each hand should be evenly balanced, and the right and left hands should be equal. I photographed my fingertips one hand at a time, with my fingers spread out so that *the thumbs were crossed*.

Photograph 2 was taken on a Friday afternoon after having the sixteenth radiation treatment. The radiation therapy was necessary to immobilize any cancer cells that might have been remaining in my breast, but the radiation also did slight physical damage to my entire energy system.[7] Since my body type is extremely sensitive to toxins of any kind, even the small dose of radiation used to kill the tiny cancers forming in the ducts of my breast gave me a very slight case of radiation sickness, and I felt as if the flow of life energy in my body got depleted and warped.

The energy is stronger in the left hand in photograph 2 even though the rays are erratic in all but the little finger. (The short black line above the middle finger of the left hand was my fingernail touching the photographic paper.)

After taking the picture, I listened to a Bernie Siegel affirmation tape and found deep psychic pain surfacing. Tears flooded my soul, and I took photograph 3 of the energy pattern while I was crying, thinking that because I was in so much emotional pain, this photo would look much worse than the one taken fifty minutes earlier.

To my surprise, photograph 3 showed intact coronas on all fingers and richer, fuller ray patterns on all of the fingers of the left hand and on two of the fingers on the right hand as well, even though the energy was still not distributed evenly.[8]

---

[7] The ionization from radiation changes the atoms of living tissue so that they lose vital substances, killing or damaging body cells. Hardest hit are the blood-forming organs like bone marrow, the lining of the gastrointestinal tract, and skin (*The World Book Encyclopedia,* 1977, s.v. "radiation").

[8]A study set up to examine the relationship between psychology and the HIV virus tested the level of killer cell activity in the bloodstream after a short-term state of sadness. The

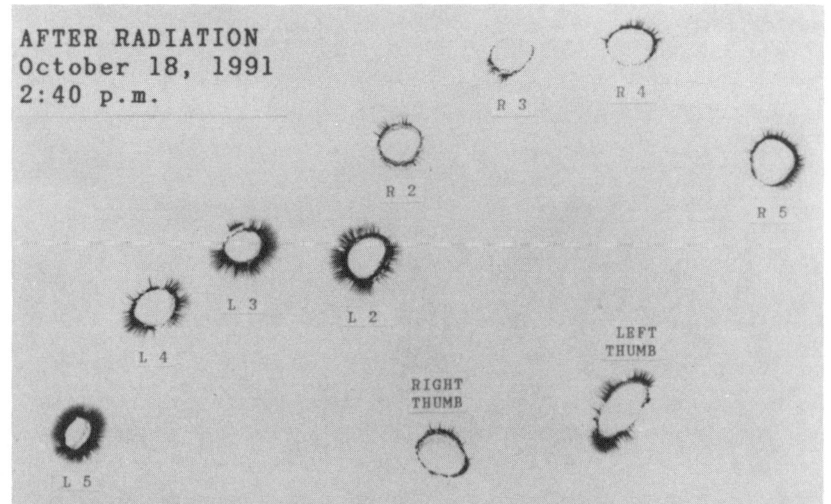

PHOTOGRAPH 2

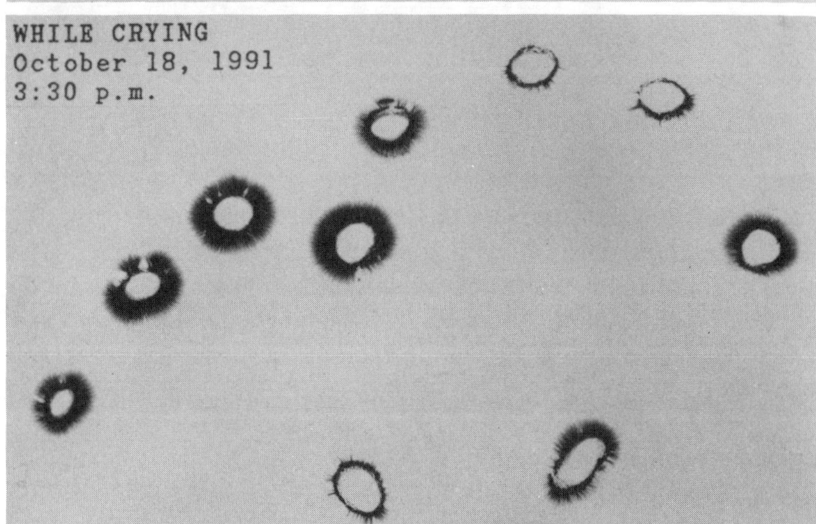

PHOTOGRAPH 3

researchers expected to find a drop in killer cell activity; however, to their surprise, during intense sad feelings, not only did the number of killer cells *increase,* but the killer cells also functioned more efficiently than they did during a neutral state, as if coming to the rescue of the sad person. The scientists then tested the effect of a short-term state of happiness and found that killer cell activity was again heightened, similar to the results during the sad state (Kemeny, 1993). This study backs up the hypothesis suggested by Kirlian photographs 2 and 3, which show an increase of energy while crying, expressing the emotion of sorrow. "It's possible that the experience of feelings per se, whether they're happy or sad, is healthy psychologically and may even be healthy physiologically" (Kemeny, p. 199). Sadness is a feeling that is often resisted because people are afraid of emotional pain; however, sadness is a normal, natural response to the loss of something of value in one's life, and expressing this feeling by having "a good cry" is therapeutic.

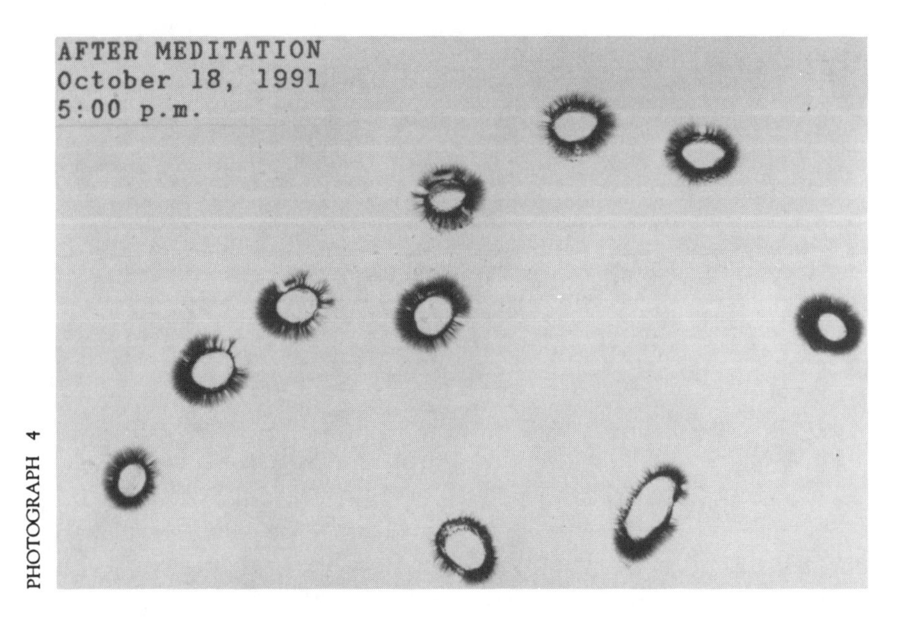

PHOTOGRAPH 4

After more crying, I meditated and felt the depression ease up, leaving me feeling more centered, with a greater feeling of internal peace.

Photograph 4, taken after this meditation, shows an energy pattern that is more balanced among the ten fingers.[9]

The following day, my physical energy level was very low as the effects of the radiation continued to cause fatigue, and photograph 5 shows very few rays of light, even though the pattern is fairly well balanced.

Photograph 6 was taken several hours after an osteopathic manipulation treatment and shows much stronger sparks than photograph 5, taken about six hours earlier.

---

[9]Studies of Kirlian photography found that the pattern around a fingertip print increased and became brighter with meditation and Yogic breathing exercises, while states of tension or emotional excitement produced a blotchy pattern (Moss, T. & Johnson, 1974). These findings correlate with studies done by Herbert Benson (1975), which show that chronic arousal of the sympathetic nervous system, producing a "fight-or-flight response," eventually leads to general damage in the physical body and cardiovascular disease in particular. States of meditation, on the other hand, activate the parasympathetic nervous system and produce what Benson calls the "Relaxation Response." This state of deep relaxation quickly decreases oxygen consumption to a level that is actually lower than the level during sleep; producing this Relaxation Response balances the body, improving physical health (Benson). States of relaxation show up in Kirlian photographs as having more brilliant, wider coronas (Moss, T. & Johnson, 1974). Perhaps cultivating the Relaxation Response gives the muscles and internal organs a chance to rest so that the life energy of the body can be channeled

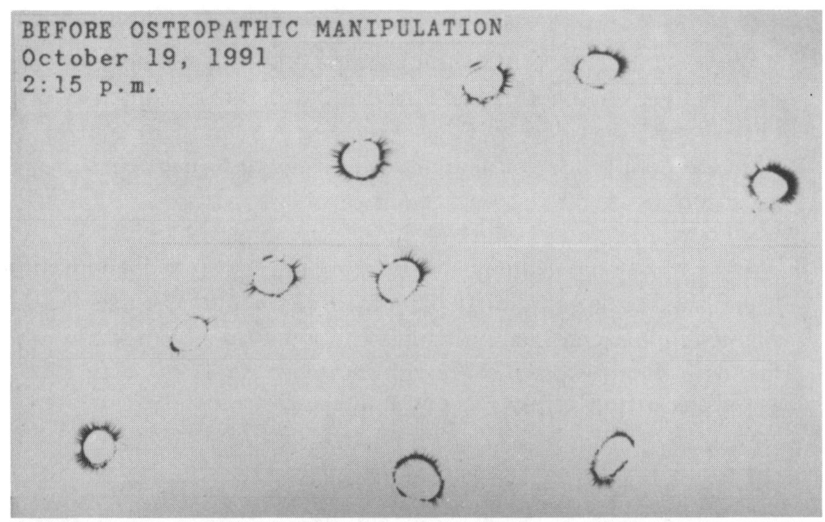

PHOTOGRAPH 5

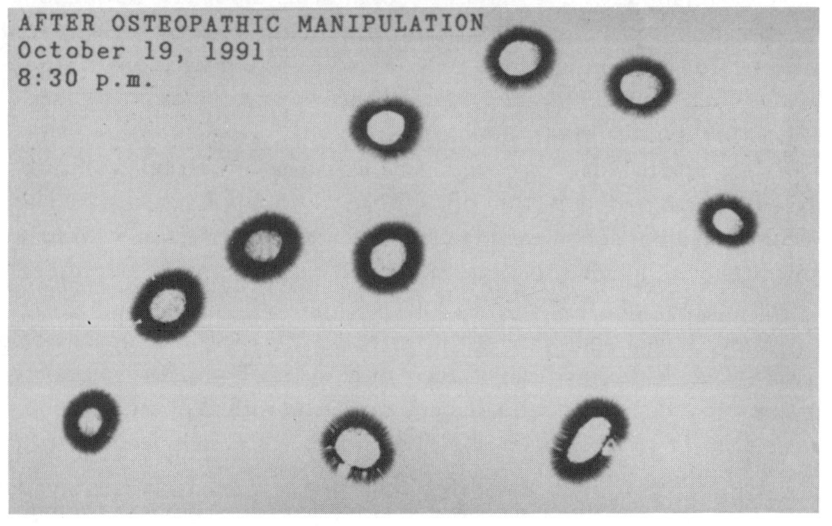

PHOTOGRAPH 6

> *November 7, 1991, Journal:*
>
> Listening to the cassette tape by Dr. William A. McCarey tonight about his book *Healing Miracles: Using Your Body*

---

into the systems of repair in the body, to correct any faulty DNA, and to energize the immune system to recognize and destroy any undesirable cells.

Energies (1988) put osteopathic manipulation treatments into perspective. I had not realized that they were the treatment of choice recommended by Edgar Cayce, a psychic who could describe people's physical ailments and suggest treatment for their cure.

I wholeheartedly believe McGarey's concept that an injured part of the body disrupts the flow of energy in the entire organism, resulting in other parts malfunctioning. The osteopathic craniosacral manipulation treatments I get from Marie feel as if they restore the flow of energy in my body, letting my body heal all parts of my system at once.

I felt very grateful for Marie's osteopathic manipulation throughout the radiation treatments and was surprised that a therapy I had never even heard of before getting cancer could restore my feeling of well-being after feeling so much fatigue and depression from the side effects of the radiation treatments.

The combination of osteopathic manipulation and group meditation gave the most perfect picture of all those we took, photograph 7. The coronas on all fingers were intact; the ray patterns were full, with only an occasional lighter spot, and the fingerprints were fully developed on all but the little finger of the left hand.

After I finished all forty radiation treatments by Thanksgiving and had two weeks for my body to recover its strength, my Kirlian photograph showed strong, well-balanced rays. The Kirlian photo taken on December 11 looked much like photograph 7 and has been omitted to avoid repetition.

The first chemotherapy session fell on Friday the 13th of December, and after taking the tranquilizer prescribed by the oncologist the previous night, the pattern on my right hand started to warp. The photograph taken on the morning of December 13 was quite unbalanced, and after having chemotherapy at 9:30 A.M., photograph 8, taken as soon as I got home after chemotherapy, showed an erratic pattern of sparks.

Marie came to my home on the afternoon of December 13 and gave me an osteopathic manipulation treatment, and I vomited periodically from 3:00 P.M. on that Friday until 3:00 A.M. the next morning.

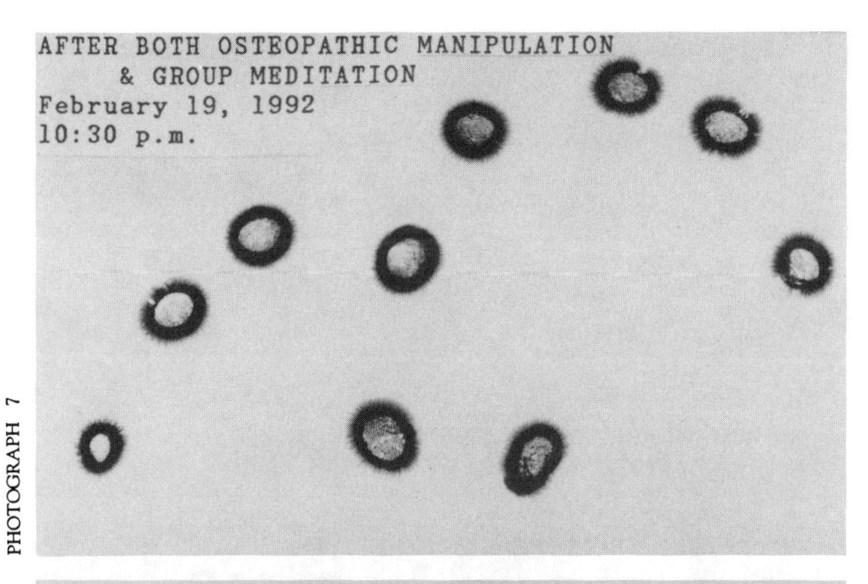

AFTER BOTH OSTEOPATHIC MANIPULATION
& GROUP MEDITATION
February 19, 1992
10:30 p.m.

PHOTOGRAPH 7

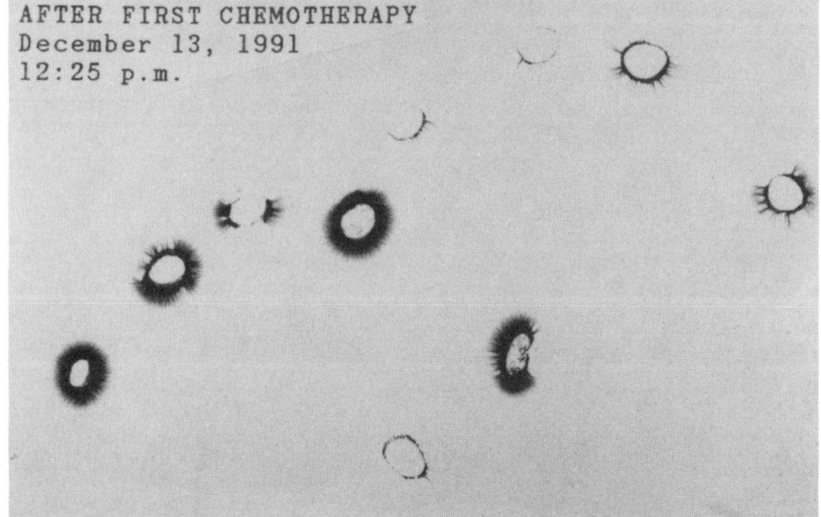

AFTER FIRST CHEMOTHERAPY
December 13, 1991
12:25 p.m.

PHOTOGRAPH 8

Photograph 9, taken at 9:30 A.M. on December 14, showed a better distribution of sparks than the picture taken right after the chemotherapy, photograph 8.

On the evening of December 14, when my friends finally located some of "the forbidden medicine," after two puffs of marijuana and meditation with the group holding hands in a circle with me, I started to feel better again. Photograph 10, taken that evening, was nearly

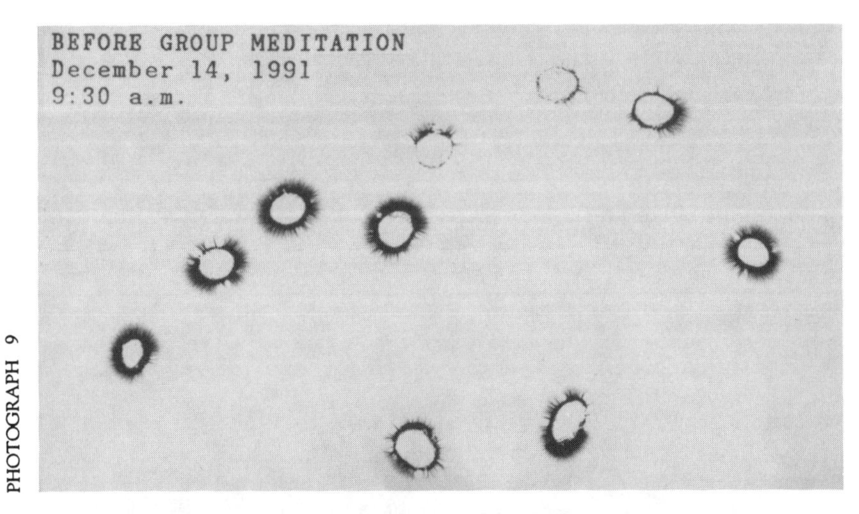

PHOTOGRAPH 9

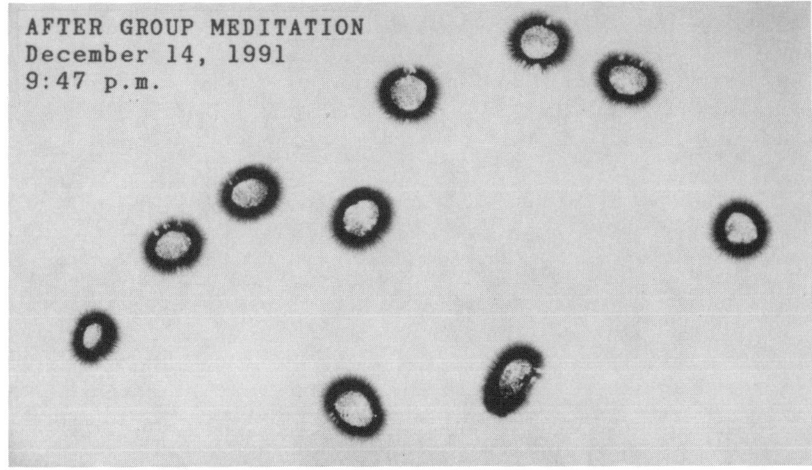

PHOTOGRAPH 10

perfect, showing full coronas, ample rays, and fingerprints on all ten fingers.[10]

Another intervention I found helpful to eliminate the toxins from my system was sweating out the impurities in the sauna. Photograph

---

[10]Research with Kirlian photography has shown that using marijuana usually produces coronas that are brighter and wider; however, in some people the coronas are unchanged, and in other people the coronas get smaller. Alcohol intoxication usually produces brilliant, wide coronas (Moss, T. & Johnson, 1974), which may be the reason that a person drinks in the first place—to feel good, to get a "high." However, a deep state of relaxation, which comes from meditation practice, has the same bright, wide coronas without poisoning the body with alcohol in the process. "Intoxication" contains the words "in-toxi . . . n": it involves *toxins* being put *in* the body.

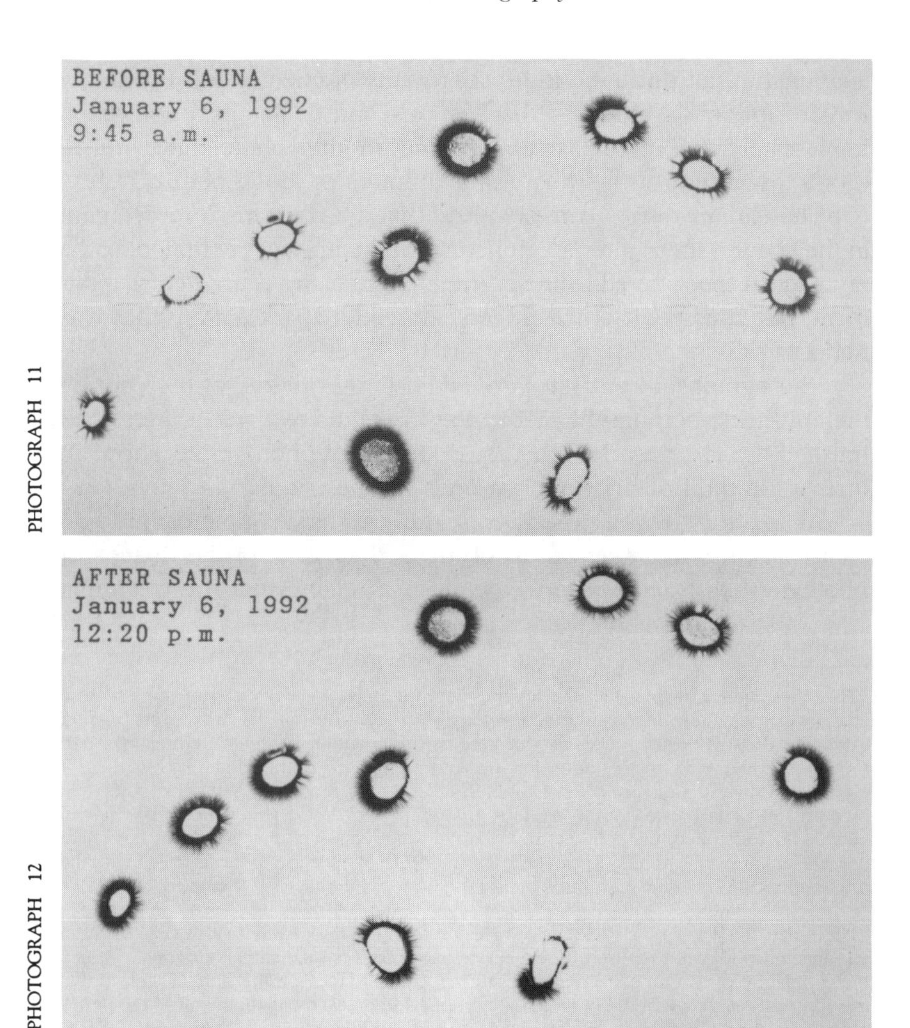

PHOTOGRAPH 11

PHOTOGRAPH 12

11 was taken three days after the second round of chemotherapy, when I was feeling a lot of nausea and depression, and the sparks are particularly weak in the left hand.

After sweating and meditating in the sauna, I felt less nausea and depression, and even though the ray patterns are not completely filled out, the overall energy pattern is stronger and more balanced between the right and left hands in photograph 12 than in the previous photo, taken before the sauna.

To add to my collection of methods I could use to help alleviate the unwanted side effects of chemotherapy, I went for weekly acupuncture all throughout the six rounds of chemo, and I took my Kirlian apparatus along to the sessions and took one photograph immediately before each treatment and another photo right after it. (For a discussion of the theory of acupuncture, see Appendix D.)

The Kirlian results consistently showed a dramatic improvement in the spark pattern after acupuncture. Sometimes a few fingers had a ray pattern that was still slightly irregular after the acupuncture treatment, but usually all of the fingers showed a spark pattern that was rich and full.[11]

Photographs 13 and 14 were taken in the middle of the chemotherapy, and photographs 15 and 16 were taken two weeks after I had finished the last round of chemo. Both sets of photographs show an increase in the brilliance and evenness of the coronas and rays.[12]

As is often true in experiments, the greatest learning came in an area I had not even set out to observe. Out of all of the dozens of photographs of acupuncture, the pictures taken after the treatment

---

[11]These results are in line with the independent findings of other experimenters that an acupuncture treatment makes the corona of the Kirlian print wider, larger, and brighter (Moss, T., 1979). The differences between the before and after photos were no surprise to me since I had read that "stimulation of an acupuncture point, either by needles or electricity, causes definite, reliable changes in the emanations around a finger pad, as seen by means of Kirlian photography" (Johnson, 1975, p. x). When a larger area of the human body is photographed, the light that shows up in Kirlian photography is more intense right over the acupuncture points (Krippner & Rubin, 1974). Stimulation of these points with the tiny acupuncture needles seems to increase the flow of "life juice" in the body.

[12]Dr. Robert Becker, who did extensive research trying to find a way that the electricity in the human body could be used to regenerate lost limbs (as salamanders do), also studied the correlation of acupuncture points with electrical resistance on the skin. He found that the electrical resistance of the human body dropped significantly at about half of the points on acupuncture maps. These points were in the same places on all of the test subjects. He postulated that the acupuncture meridians were "electrical conductors that carried an injury message to the brain, which responded by sending back the appropriate level of direct current to stimulate healing in the troubled area" (Becker & Selden, 1985, p. 234). If Becker's model is accurate, then acupuncture could be used in two main ways:

1. Blocking the incoming signal, which includes sending a pain message to the brain, by inserting a metal needle into the proper point to short out and stop the pain message. (The success of using acupuncture for anesthesia is well known.)
2. Boosting the outgoing signal of direct current sent to stimulate healing of the troubled area. The minute electrical impulses the body uses to send healing to a wounded area consist of direct current measured in nanoamperes (billionths of an amp) powered by microvolts (millionths of a volt). Since the resistance along any transmission cable makes a current get weaker with distance, engineers build in amplifiers along power lines to boost the signal back up to its proper strength. Becker theorized that the acupuncture points were like miniature amplifiers, and the Chinese had long ago discovered how to insert needles into these points to balance and harmonize the flow of energy in the body.

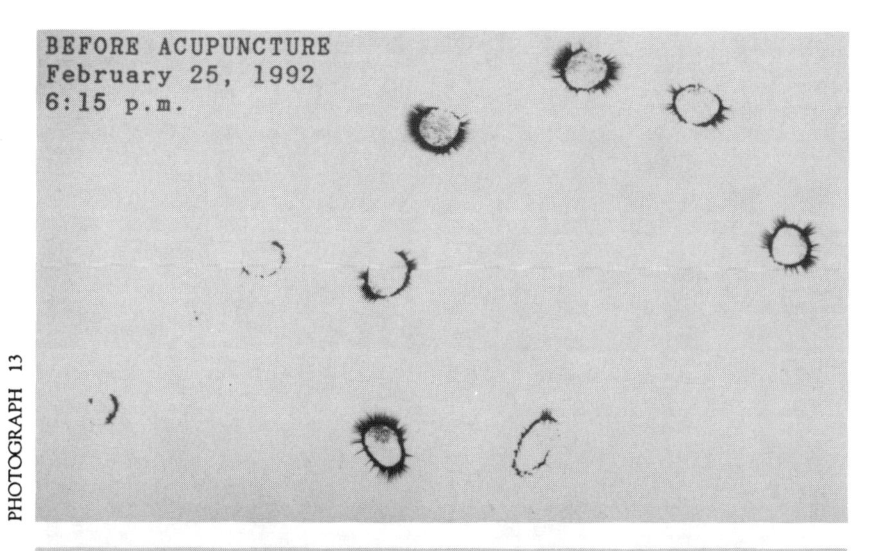

PHOTOGRAPH 13

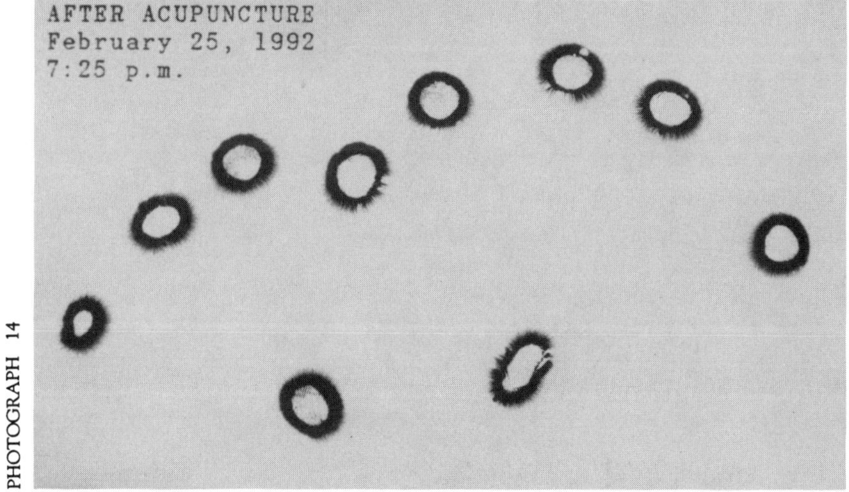

PHOTOGRAPH 14

always had stronger sparks than the pictures taken before, *except for the set taken on April 21,* (photographs 17 and 18) where the photograph afterwards showed weaker rays and breaks in the coronas in several fingers on the left hand.

Although after that treatment, both Suki and I had thought the sparks were darker, the photograph showed that they were lighter. (My subjective observation was *not* confirmed by the photograph.)

When I got the results of these photographs 17 and 18, at first I thought that I had mislabeled the before and after pictures. But then I

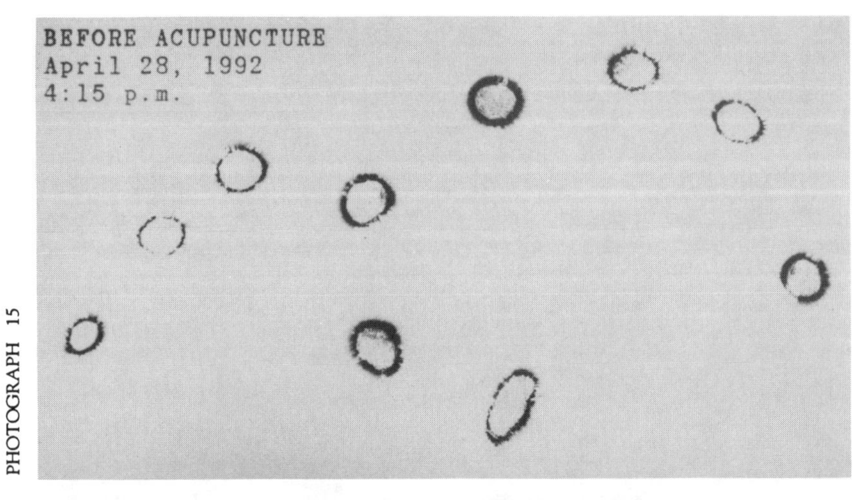

PHOTOGRAPH 15

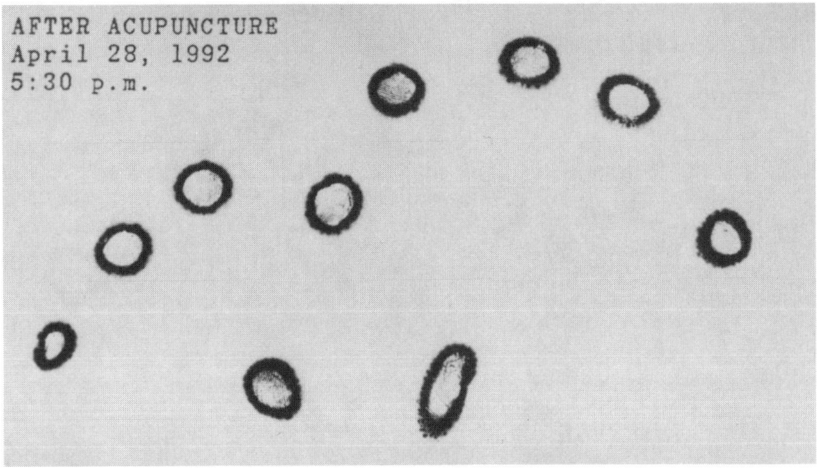

AFTER ACUPUNCTURE
April 28, 1992
5:30 p.m.

PHOTOGRAPH 16

realized that we labeled each photo so carefully that a mistake like that was not possible.

During all of the other acupuncture treatments, I had meditated during the time while the needles were in; however, on April 21, the subject of suicide had come up. The suicides of Peter and Marcela were still fresh in my mind, so instead of meditating during the treatment, I spent the whole hour while the acupuncture needles were in my skin talking with Suki about the emotional aspects of suicide and its risk assessment, from both my professional training and my personal experience in this difficult subject. I was hoping to

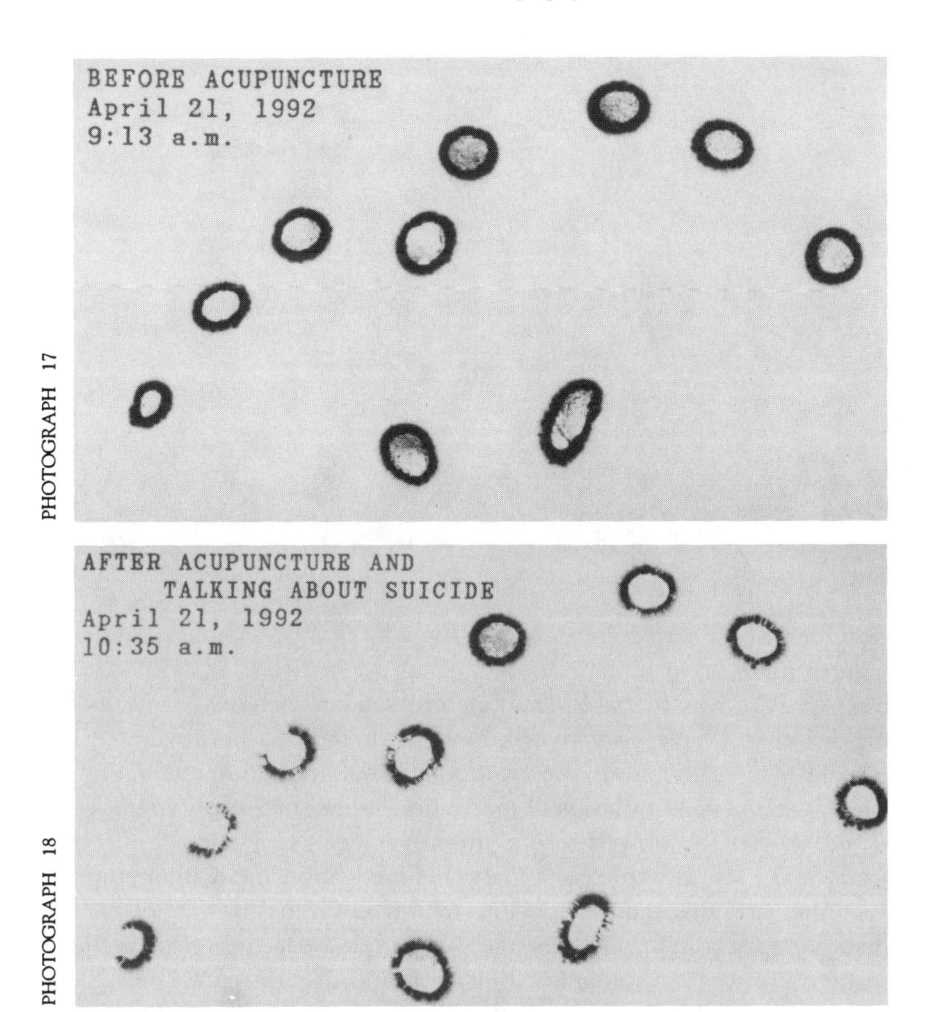

PHOTOGRAPH 17

PHOTOGRAPH 18

share some knowledge that might help prevent another tragedy like the deaths of Peter and Marcela.

The sparks in the photograph taken after concentrating on this negative, destructive topic for an hour appeared weaker and disrupted in the left hand. Several days later, the ex-wife of one of Suki's clients slit her throat and then called the ex-husband, informing him of what she had done. Fortunately, he got help to her before she bled to death, and Suki was able to be more sensitive to the emotional dynamics of his situation because of our talk. I wondered why the topic of suicide had come up in my session with Suki the week before,

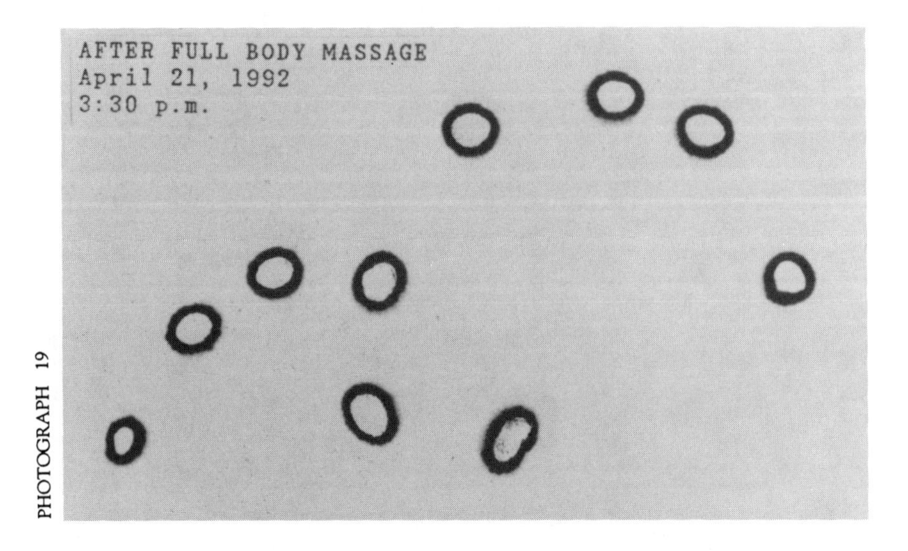

PHOTOGRAPH 19

whether it was pure coincidence or some kind of premonition on her part or mine.

April 21 was the only acupuncture session in which I did *not* meditate, so I have no way of knowing whether the improvement I usually saw came from the acupuncture or from the meditation. However, the pictures I took of meditation alone, such as photographs 3 and 4, usually showed distinct improvement, but not as perfect a pattern as I saw after the combination of meditation and acupuncture.

After finishing the acupuncture treatment on that day, April 21, I had gone for a full body massage, which felt great and restored the pattern of rays on photograph 19 to a balanced state again.

Another result really surprised me because it was the total opposite of what I had expected. The very first photograph we took once the apparatus was working was done right before the spiritual "attunement" by Gloria Karpinski on October 2, 1991. I had expected to see stronger sparks after the attunement because I felt so uplifted by the encouragement its content gave me; but instead of improving after the attunement, I could see *no sparks coming out of my fingertips at all.* I could not believe my eyes and thought that the battery in the Kirlian apparatus was dead, so we took another photo with a fresh battery, which still did not produce any sparks. I wondered what was wrong and took a third picture, unsuccessfully trying to get some sparks to show up from the ends of my fingers.

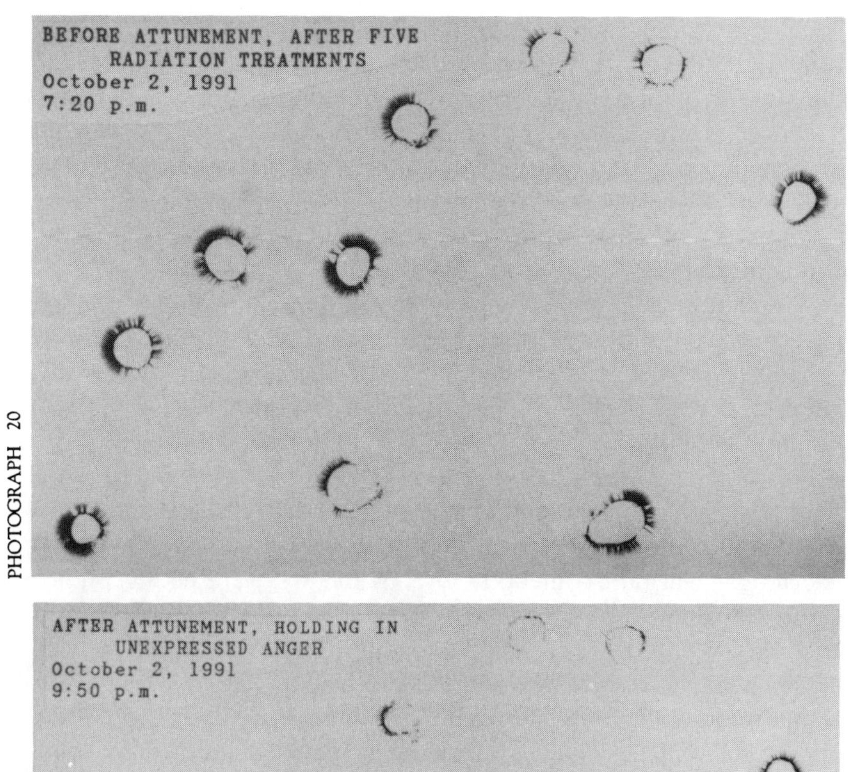

PHOTOGRAPH 20

PHOTOGRAPH 21

The developed photos showed that the sparks in photograph 20 taken before the attunement were a bit weak and spotty, but the sparks in photograph 21 taken afterwards were much worse, almost nonexistent.

Before talking with Gloria, I had asked Soli to go to take the Kirlian photograph in silence after the attunement so I could stay in the state of mind I would be in when the attunement was finished. Soli forgot the request and was making friendly chatter on the way out to the darkroom. I felt angry that he was talking to me, but instead of answering his questions or putting my finger to my lips to remind him to hush, I said nothing and tried to maintain the "spell" of the attunement, as my irritation increased with every step.

The attunement had filled me with joy and energy and had given me a spiritual "high" of hope and faith that I could survive the cancer. However, the absolute worst of all 140 Kirlian pictures was photograph 21, taken right after the attunement, when I was mad at Soli because I thought he had spoiled my experiment, but I was not expressing that anger.

I HATE that familiar battle within when I am offended and angry about some little thing but fear that if I express that anger, I will be rejected or will hurt somebody else. While we were taking photograph 21, one voice inside was steamed at Soli for talking, and another voice said I was being very silly to get mad at such a nice person who was obviously *trying* to be of help to me. The angry voice within retorted that Soli had ruined my attempt to observe the effect of "spiritual enlightenment" on my spark pattern, and a third voice within criticized me for getting angry in the first place and for getting so worked up over nothing.

I felt the physiological effects of this silent inner battle: slight shaking in my body, a heavy knot in my stomach, and a feeling of pressure in my head, as if my brain wanted to explode.[13]

When I heard a tape of Carl and Stephanie Simonton (1975) saying that the cancer personality looks to the needs of others so much that they express positive emotions but not negative ones and are "too good to be true," I did not want to think that pattern applied to me. Also, LeShan's work on the emotional patterns of the 500

---

[13]Researchers Moss and Johnson noticed consistent changes in photographs taken while a subject was in a state of anger and show an example on p. 79 in their book (Moss, T., 1979). They were photographing only the index finger of the hand and found that the spark pattern around the non-angry fingertip was wider and fuller than the one around the angry fingertip. Since they were doing color photography, they also noticed a red blotch in the angry Kirlian photograph, a finding they found consistently when the subject was in the emotional state of anger. (My black and white photographic paper could not pick up red light.) Perhaps the folk description of being angry as "seeing red" has some literal truth to it.

cancer patients he researched found that cancer victims have a "bottled-up" quality to their emotional lives (1977, p. 57); however, I did not want to think that rule applied to me either.

But *something* happened between photograph 20 and 21 that severely cut down the amount of vitality in my body. And I know from life experience that marital stress always drains my energy. My silence, keeping my feelings to myself, increased that stress.

As I studied photograph 21, I noticed that the left hand ring finger, where I wore my wedding band, was almost entirely absent from the photograph. All of the other nine fingers showed up at least a little bit. Our culture wears the wedding ring on this particular finger because legend says a vein from this finger goes directly to the heart. The fourth finger is also considered to represent creativity, and the union of two partners in marriage often results in the creation of additional people. I wondered whether holding in my anger so I would not offend my partner for an innocent wrongdoing might hurt my heart and injure my own creativity.

Other Kirlian researchers found that when a person is in a state of emotional withdrawal, the energy body retreats from the extremities of the body, producing no picture (Snellgrove & Snellgrove, 1979).[14]

Anger is a natural response when another person blocks an intended goal, even if done unintentionally like Soli had done; however, a healthy emotional response would have been to address the issue in an appropriate, assertive manner instead of boiling in silence as was my customary pattern.[15]

---

[14]This finding is consistent with the psychological theory of Sigmund Freud that in the states of both mourning and melancholia (depression), loss of a love-object makes the ego gradually withdraw its emotional energy, its libido, from the outer world (Freud, 1959).

[15]Some people have a pattern of inappropriately expressing constant hostility to others, part of a personality pattern labeled "Type A." However, over-expressing anger was hardly my problem. When a strong emotion like anger is expressed appropriately, it can then be released and forgotten. But anger held inside can turn into hostility, which then affects the heart. Original studies had indicated that the hard-driving, impatient, hostile, fast moving Type A pattern behavior was a greater predictor of heart disease than the more easy-going Type B pattern; however, a later study showed the Type A pattern had a higher *long-term* survival rate than Type B (Ragland & Brand, 1988). A further study isolated the variables in the Type A pattern and found that only the component of *hostility* in this pattern increased the risk for illness. "What is more, high hostility scores predicted not just myocardial infarction and death from heart disease, they also predicted increased risk of death from cancer and all other causes as well" (Kabat-Zinn, 1990, p. 212). Perhaps hostility produces biochemicals, which somehow damage the function of the immune system. A study of the

Many people consider anger and sadness to be "negative" emotions and go to great lengths to avoid facing these feelings; however, the Kirlian photographs suggest that feelings need expression to encourage "life juice." When I fully entered my pain in photograph 3, expressing my feelings of sadness and depression through tears of sorrow, the spark pattern got stronger. The reverse attitude of holding my feelings of anger inside in photograph 21 seemed to almost completely block the flow of sparks.

A cancerous tumor shows up on a Kirlian photograph as an area with no light coming out, no sparks at all. Perhaps cancer has no "life energy," only death energy. Witholding my emotions seemed to block the flow of life energy in my body in a pattern that showed up looking quite similar to the pattern cancer has: no life juice.

Since I first saw these photographs, I have consciously worked at changing this "silent boiler" pattern of mine and have been talking with Soli and others when I feel offended by something that happens. Because of this change, I feel more emotionally connected to my world. I feel much better after an interaction in which I express myself, and then I can more quickly get the whole situation into perspective and let go of my irritation.

Now when I get upset about some little thing Soli does and tell him, he reports his view of the situation and then says to me, "I see you are expressing your feelings, and I am too!"

---

immune response of Type A's versus Type B's found that while the immune activity of the Type B personalities recovered within 15 minutes after the stress of mental mathematical subtraction, the immune response of the Type A as measured by lymphocyte proliferation was still low one hour after the stressor. The Type A's did not have the flexibility to recuperate from the onslaught of stress as quickly as the Type B's could. However, the immune response of both Type A and Type B subjects practicing Transcendental Meditation evidenced a Type B immunoreactive pattern under conditions of stress and rest (Blasdell, 1989).

CHAPTER 15

Healers

The Kirlian photos helped me deal with my curiosity about the question, "WHAT HEALS?" The topic of healing is so broad and multifaceted that a person can view only one angle at a time, walking around the mystery of healing and trying to understand what has happened when healing comes. Many healing sessions were necessary in my journey through cancer, because cancer itself is a complex illness, and I could only release one layer at a time of difficulty in the physical, emotional, or mental area. In addition, I also needed continued healing from the side effects of the cancer treatments themselves.

Healing came to me in many forms, and of the many healers who came into my life during the journey through cancer, two particular teachers impacted the way I viewed the relationship between my body and my mind and helped me heal, both emotionally and physically. This chapter relates the interactions with these two people.

Both of these healers were engineers who were familiar with the research of the late Marcel Vogel, a senior IBM scientist who was a prolific inventor and made many prominent contributions to technology, including developing the magnetic disc coating for IBM computer hard drive magnetic storage devices and developing the phosphors that made color TV possible.

A month after diagnosis of cancer, in September, 1991, when my friend Sara first suggested that I see an experienced healer named Tom Milliren, a retired engineer in his sixties who calls himself a "psychic researcher," I refused. Sara had said that Tom did exorcism, a word which brought up many negative connotations in my mind. *I did not feel I had any negative spirits in me; I just had cancer.* Sara and

I had a little tiff over her insistence that he could help me and my refusal to comply with a treatment I did not feel I needed.

When Tom was passing through our area again in March of 1992, Sara again suggested I make an appointment with him, and she mentioned that he also did emotional life regressions to heal past emotional trauma. I still had major doubts about whether this person could do anything to help me, but I wanted my friendship with Sara back, so I decided to give Tom a try. I *knew* I had trauma locked up in my body, "issues in the tissues," and I wanted this trauma totally healed. I wanted to unlock and clean out the buried emotions of shame, anxiety, mistrust, and fear which I felt in my energy field.

I went for a healing session with Tom on March 15, 1992, between the fourth and fifth rounds of chemotherapy. At that time, both my body and spirit were suffering from the onslaught of chemical poisoning from the drugs used to kill the cancer. To my surprise, I found Tom to be a gentle, caring elderly man, not at all like my stereotype of a hellfire and brimstone "exorcist."[1]

When Tom began working with me, he prayed for divine protection and guidance, and I added a prayer for my own healing and protection, setting my intention to release whatever negative patterns were locked up inside of me so I could completely heal. Tom prayed for all imbalances of energy to be released and worked from the top of my head on down through my body to get my energy balanced and centered. Tom sensed a spot in my belly that was like a leak in my energy field and wondered whether this spot might reflect some pain I brought in with me at birth. The following journal entry shows the results of my internal search for the source of this "energy leak":

---

[1]Tom is a member of the American Society of Dowsers and has written the following two books, which he has published himself:

Milliren, T. J. (1990). *Learning the Art of Dowsing/Divining: A Practical Approach for the Student* (2nd ed.).

Milliren, T. J. (1993). *Noxious (Geopathic) Fields Are Damaging to Your Health* (2nd rev.).

Both books are available at the following address:

American Society of Dowsers Bookstore
101 Railroad Street
St. Johnsbury, VT 05819

March 15, 1992, Journal:

As Tom worked on me, I visualized the pain I had chosen to bring into the world with me as a cross I had lugged through the birth canal and had been dragging around with me all my life. I released this pain, which I saw had been a magnet pulling me back home, into death, to go home to God.

As the pain was finally released, I wept deep tears because I realized that since I had made the decision to release that pain, I would have to finish out my natural life span here. No cancer shortcuts to heaven!

Then Tom asked if I would like to release the rest of the emotional traumas I was carrying through an emotional regression. Cancer taught me that life is too short to spend it dragging all my pain around, so I wholeheartedly exclaimed, "Yes!"

Tom had me lie down and knelt beside me, holding my hand as I closed my eyes and relaxed into a state of deep meditation. Then Tom said, "Imagine the most potent emotional trauma in your life and tell me what age you were."

I connected with an incident that had happened when I was about nine years old, and tears started to flow. Tom asked me to describe the images that were producing the tears. The following trauma ran through my mind:

A close friend and I were playing up in the hay loft of my grandfather's barn, when suddenly he grabbed me and held my left arm. He put his hand on my left breast (the one which got the cancer) and asked, "What 'cha got there?"

I honestly did not know what I had there, so with my right arm I punched him in the chest and yelled, "Same thing you got!" He then grabbed my other arm and held both wrists in one hand while he put his other hand into my crotch. I felt

> a sexual tingle, a strange, unknown, frightful feeling. I did not know what was happening nor what he wanted from me. I did not know what sex was at the time, but I felt very scared and angry.
>
> I fought him and told my four-year-old sister who was up in the hay loft with us, "Bring me the pitchfork!" She did not want to, so I threatened that I would mess up her room every day for a year if she didn't. She then changed her mind, and when she brought the pitchfork to me, he let me go.
>
> Afterwards, I went and sat in the car for a long time, feeling confused, frightened, vulnerable, and dirty.

I felt so ashamed about the incident that I never told my mother or anyone else what had happened. I could not stop sobbing as I told the story of being molested and felt the pain of that trauma and how it had made trusting men difficult my whole life.

Tom said to me, "He did not penetrate you. You resisted him. He was experimenting sexually and did not mean to hurt you. Can you forgive him totally and completely, without reservation?"

I cried some more, and I knew in my heart that what Tom said was true. My friend's intention had not been to hurt me; he was curious about sex, and his intention had been to try out having sex with somebody. But he violated my free will by trying to force me into sex with him, and he frightened and confused me terribly in the process. *He deserved to be punished!*

I felt the full weight of the debt he owed me, of all the emotional trauma that had resulted in my life from those five minutes, which he probably did not even remember. I decided to release him from having to pay that debt, and more importantly, to release myself from having to collect it. I forgave him, totally and completely, without reservation. I deleted the debt from my emotional balance sheet.

Tom then inquired, "Can you forgive yourself?" My first reaction was feeling that I had not done anything wrong, and the other person involved deserved all of the blame for what had happened. Yet I felt so much shame from being involved in the incident that my feelings were acting as if I myself had done something wrong. I realized I had been incorrectly blaming myself in some way. I was also carrying around a "guilt trip" for imagining that I had been sexually defiled, and I felt dirty because a male had touched me in a private place. I felt ashamed for having been a part of the abuse, even though I did not instigate it.

The guilt I was carrying around was false guilt, and it pulled down my self-esteem. I needed to let go of this false guilt by forgiving myself for the self-blame I had fabricated. I forgave myself, totally and completely, without reservation. As I let go of the lie that "I was bad" because somebody had molested me, I felt sweet, exquisite relief pour into my heart from releasing the false guilt. And once I forgave myself, my low self-esteem was unhooked from the burden of shame weighing it down and could bounce back up to normal.[2]

Tom began counting down year by year from nine and asked me to tell whatever other emotional traumas came up. Tears welled up again when he got to age five, remembering Mom whipping me with a belt for spilling strawberries on my yellow dress on Sunday morning right before going to church. I wept as I told the story.

My depth perception is defective and was only 10% of normal at that time, and I had misjudged the position of the strawberries. I felt the punishment was so unfair.

When I finished verbalizing the trauma, Tom said, "Your mother was confused. She had her own problems and did not know how to treat you properly. Can you forgive her, completely and unreservedly?" I forgave her from my heart. "Can you forgive yourself?" I did.

He then counted down further, and more tears flowed when he got to age one. I told him of connecting with a feeling I had as a little baby that my father did not care about me.

Tom said, "Men often do not know how to express their feelings. Your father does love you. Can you forgive him? Can you forgive yourself?" Again I forgave and released the feelings from the past, adding that I had really felt my father's love when I got older.

---

[2]Forgiveness is *not* the appropriate response to sexual abuse if the abuse is ongoing, as in a child custody case where a parent is still actively sexually abusing the child. The first step in healing is to stop the abuse so the child is not re-traumatized. The long healing journey of getting treatment for the child and also for the perpetrator can only begin once the abuse has been halted. Eventually, the child will need to forgive the perpetrator and the internalized false sense of guilt before the wound can truly heal. Most perpetrators are themselves childhood victims of physical and/or sexual abuse, and the perpetrator will need to ask the forgiveness of the victim and will also need to learn to forgive himself or herself before the pattern of abuse can be broken. Otherwise, the perpetrator will just find another victim to abuse, and that victim may then become a perpetrator.

The pain that I had lugged through the birth canal was a deep sense of not being worthy or loved, and because those feelings were already in me when I came into the world, I interpreted normal parental activities like my father being very involved in his work and my mother rigorously disciplining me as evidence that I was not loved. And adolescent sexual experimentation had felt like a traumatic invasion of my feminine body. These very wounds were the catalysts that propelled me into a higher level of consciousness, to seek the very Source of Love, which could heal my heart.

Tom asked if I saw any other traumas, and two incidents from after nine years old came up; however, these memories surfaced without a strong emotional charge. I saw I had already forgiven much of those two issues, but he again had me go through and forgive everybody involved.

When the emotional regression was all over, I felt like a new person, internally cleansed and radiant. Tom said that purifying and centering the energy body will often produce a dramatic positive change in the physical body, and everyone there noticed that my face looked different afterwards, all shining and happy. That night I had the following dream:

March 16, 1992, Dream:

I am with the person who molested me and others in a large house. I notice that he looks young, slim, and handsome, and I am interacting freely with him. I see that the reason he was tempted sexually with me was because we took naps together a lot, cuddling together in the same bed. I feel my love for him again.

When I awoke from this dream, I was amazed to realize that the pain from being molested was *gone.* I still had the memory of the incident, but it no longer carried a negative charge: no pain was attached to that memory. I had released it all.

The dream reflected the transformation that happened in me and revealed the basic truth about the relationship this person and I had: we had been close. Although we never took actual naps together, the dream correctly pictured our unconscious affinity, the kinship we felt with each other. The dream enabled me to feel my love for him again and his for me.

I later wrote to him, telling him of the emotional regression, the dream, and expressing my forgiveness directly to him. I was surprised to find myself laughing as I wrote out the account of being molested. I thought it was funny that I really did not know what my breasts or my crotch were for and thought they were the same as his. Also, I found humorous the fact that the worst thing I could think of to coerce my little sister into bringing me the pitchfork was to threaten to mess up her room, since I myself had one of the world's messiest rooms as a child. Being able to laugh about what had happened was a true sign that the wound had healed.

The second healer who touched my life deeply was Robert Fritchie, a professional engineer/consultant who had worked with Marcel Vogel. Fritchie and Vogel became interested in the therapeutic use of quartz crystal because of its widespread use in science.[3]

Vogel and Fritchie spent many hours after their regular scientific jobs, working far into the night, researching the geometric forms of quartz crystal that could be used for physical healing.[4] Although they found a way to focus energy for physical healing using a quartz crystal, they then discovered that they could get the same results without using a crystal by simply setting the intention to channel spiritual love and healing to the recipient.

Furthermore, Fritchie found that when a group of people gathered with the intention to use their collective energy for healing, the process of repair in the physical body could be accelerated exponentially by the group dynamics. This discovery of Vogel and Fritchie formed the basis for the healing meditation group I had been attending.

I got many ideas on how I could heal myself from Fritchie's unpublished manuscript my friend Starr brought me, titled *A Path to Self-Healing Using God's Love and Crystals* (1987). Fritchie taught that healing with divine love was releasing the energy blockages a person

---

[3]Quartz crystal, silicon dioxide, is the basis of many modern inventions of technology, including quartz watches, guidance systems for space missiles, radio receivers, and sound systems. The computer disc memory systems that Vogel had discovered and patented are a technology based on crystalline forms, and the extremely important timing devices in the computers are crystals.

[4]Vogel cut crystals to specific dimensions with a termination on each end, and he and Fritchie found they could channel healing energy through these specialized crystals into a sick person asking for healing. Both they and the people they worked with were surprised by the amount of improvement reported, and Fritchie (1987) wrote about their discoveries.

had to his or her own natural healing process; he said the healing always came from God, but having another person be a "server" facilitated and sped up the process.

When Starr told me that Bob Fritchie was coming to Mansfield, Massachusetts, for a workshop on "The Group Healing Process" on March 25, I felt excited to have a chance to work with him in person. When we arrived at the workshop late, Fritchie was talking about perception and intuition, about trusting an inner connection that simply *knows* things. I immediately felt that Fritchie was speaking my language and listened intently to his teaching. He said all healing came from the connection to God inside of a person, and anyone in the role of "healer" was just an assistant to serve that process.

He claimed the key to group work was *intention* and that exactly what was done mattered little, other than having the pure intent to be of help to other people. He felt the process could also be enhanced by putting a hand on the shoulder of the person receiving healing, to ground the energy exchange in human contact.

Fritchie also taught that the key to the effectiveness of the healing work was forgiving everybody that had ever "done you dirty" and forgiving oneself, the same principle Tom used in his work and mind-body authors Bernie Siegel and Louise Hay expounded.

Fritchie told of slides Marcel Vogel had shown of liquid crystals being charged with Divine Love: they first showed the faint outline of a new form, a flash of light followed, and then the particles jumped into the new shape. Then Vogel had shown a slide of a blood cell, which when charged with Divine Love, also outlined a new form, showed a flash of light, and then jumped into the new shape. Vogel's demonstration built a powerful case for treating the entire human being as a crystalline form that could be transformed by using Divine Love to heal.

Fritchie also stated, "Your brain/blood is a liquid crystal," and he taught that a person could influence the function of the cells in the body by sending them Divine Love, which would provide the energy for the DNA of any abnormal, pathological cells to return to its normal function. I realized that I could visualize a program of total health at the cellular level in my body and use Divine Love Energy to activate that program in me.

On the first round of the group healing, Fritchie asked me to release any cancer cells that might be remaining in my body; and on the second round, he instructed me to visualize my body being in a state in which no chemical that was introduced into it could do any harm, as I still had one more chemotherapy session to go.

Fritchie's theory of the origin of cancer was that it was often the result of "emotional overload." He felt the sum total of the stressors in a person's life, whether from "programs" dumped on from oneself or from others, could impact the body, causing cells to lose their DNA memory and mutate. In addition, cells could be damaged from radiation, drug accumulations, and certain harmful electromagnetic frequencies, which upset the normal energy balance of the body. These energy impacts, if not released by some means, could become like "mud balls" in the system, causing cells to lose their reference for cell regeneration, and then those cells could mutate into cancer or a host of other diseases. Fritchie taught that focusing spiritual love with an intention to heal into those very same mutated cells could clear out the old "programs," allowing the cell to recover.

Fritchie noted that a group of symptoms signals high stress in a person: the head clogs up, the nose and eyes run, the chest feels tight, and the body has aches and pains all over (Selye, 1976). Living in a stressful environment or doing stressful work can contribute to the overload, and the end result is that the body cannot process all of the stress. He added that usually stress related energy goes into the weakest part of the body, blocking cell activity and finally contributing to cell mutation; any area that has had an injury is fair game to get attacked by the cancer. Fritchie felt that because of the emotional blockages men often have, the stress energy usually enters men through the solar plexus and resides in the lower trunk, leading to prostate cancer; women are often negatively impacted in the heart region, which then radiates into the breast or the lungs.

To prevent this negative result of stress, Fritchie recommended taking note of times when the body was going into the stress reaction and right at that moment, intervening in the stress process by breathing in spiritual energy and spiritual love to break up the overload.

Fritchie instructed the group to do deep breathing while holding our collective intention to heal. This breathing synchronized the group to program wellness back into the cells of our bodies.

After the session, I wrote in my journal:

March 26, 1992, Journal:

Never before have I felt such complete wholeness and healing, as if Lord Jesus Christ said to me, "Now is the time." I got positively charged from my own acceptance of the spiritual love of the group.

During the part about releasing all past emotional traumas, I felt my slate was already clean from the previous work with Tom.

Not even the disturbance of arriving late and Soli's leaving the group because he felt upset can take away from the wholeness I feel. I realize I am responsible to seek my own healing and cannot ask Soli to do it for me. In fact, maybe I needed to go through this threshold alone.

This morning the rash on the back of my neck was gone, and my left breast nipple was the most normal I have seen it since the surgery. I also slept straight through the night, the first time since Sunday, March 22. And I didn't even consciously work on any of those three things.

I'M HEALED!

I feel absolutely wonderful, full of love!

In my weakened condition from the side effects of chemotherapy, I felt dramatic physical healing in my body from the work I did with Fritchie and Milliren. The work Fritchie did with me was aimed at the cellular level of organization, to rewrite the instructions in the DNA of damaged cells to make healthier cells. The work Milliren did with me provided a structure through which negative emotions could be transformed. The emotional regression took me back to painful feelings and gave me a framework in which I could express those feelings directly, a new event for me, and forgiving everybody released my feelings of hurt and shame.

Sometimes forgiveness is mistakenly equated with repressing and may be the reason why the adage "forgive and forget" may be seen as too naive. The mind needs to remember events to keep a strong sense of identity. Milliren's method of healing did not lead to forgetting or repressing the incident; instead, he helped me to remember the experience and then reframed it through the eyes of deeper understanding, transforming the emotions attached to it, allowing my soul to heal and filling the wounded space with love.

I had no doubt that the healing effects of the work with these two healers was real because I felt healing changes in my physical body; but I also felt healing at the emotional, mental and soul level. Joy filled my heart, my mind was clear, and deep peace was present in my soul.

CHAPTER 16

Finishing Chemotherapy

Getting through the side effects of chemotherapy was difficult, but chemotherapy also indirectly brought me into new avenues of healing. The following dream comments on the process of regeneration that was at work in me as I went through cancer:

> *January 18, 1992, Dream (first half):*
>
> I move back to the college where I started my freshman year to take another year of study and happen into a Music Appreciation class by a former professor, Mary Oyer.
>
> She looks radiant. I know she has recently been through breast cancer and had a mastectomy. She comes and embraces me with joy, saying, "NOW YOU ARE LEARNING TO LOVE YOURSELF."

The dream begins by showing that I am in a period of intensive new learning, like starting college again. The area of study that the dream wants me to learn is not Cell Biology, Anatomy and Physiology, or Cancer Pathology, but simply *Music Appreciation,* tuning in to the beauty and joy of the music of life.

Although music was my undergraduate major, in college I focused more on perfecting my piano performance technique and learning music theory and music history than on immersing myself in gratitude and appreciation for the wonderful gift of music in the world.

The instructor in the dream was a dynamic woman who in waking life had recently been through a mastectomy; as a breast cancer survivor, she joyfully embraced me with the simple key to healing: LEARNING SELF-LOVE.

However, the process of loving my life included undergoing chemotherapy to give me a better chance at having more time on the planet, and enduring this drastic medical treatment was rough on my instinctual nature. The dream continued:

January 18, 1992, Dream (second half):

Riding along a country road in the back of the wagon, I point out animals to the others with me: an armadillo beside the road and a crow hiding in the bush. I notice a large snake in the wagon right beside me. I take it, holding it by the neck, and hold it up to others asking what I should do with it.

I decide from its markings that it is poisonous. First I think about just throwing it off the wagon, but it would have a chance to strike me on the way down. Then I hold my other hand just below its neck and ask others to cut through its body. An animal bites/saws through but leaves just the backbone. With my bare hands I break the neck of the snake and throw it out behind, with blood and body liquid all over my hands and arms. I go to wash up.

I awoke from the dream horrified at the way I had treated the snake, the symbol of healing. My ego was identified with being a "nice person," and the viciousness against the snake in the dream alarmed me. As I took this image and my fear that I was doing something wrong into prayer, I felt Lord Jesus Christ putting the clear thought into my mind:

> "You have broken the back of the poison used to heal your body. Rest now and trust in me."

When I went to the doctor on January 24 for the third chemotherapy treatment, my white blood cell count was only 3.3, too low to do chemotherapy again, so the treatment had to be postponed for a week. Since I had already gone through the emotional agony of getting ready for the session, my first reaction to this news was anger and disappointment. However, then I remembered the dream I had during the previous night:

January 24, 1992, Dream:

A Pacifica classmate and I are getting ready to go for chemotherapy. Before I finish getting ready, she drives me down a block.

I complain to her that I am disoriented and don't know how to get back. Would she please take me back? Instead, she drives me farther away. I weep because I don't have on my underwear, don't have my hair done, and don't feel prepared for the chemo. Then I say to her, "Thanks for being a good friend and putting up with the grumpiness that comes from getting chemotherapy."

I go to a place where my pottery teacher is in charge of making a movie. He explains everything well and films as we go. It's fun. Then they want me to make a human circle with my body, and they will film me rolling along like a wheel. I decide my body won't withstand that physical treatment. Can I play another part? Yes, that of a person who is dead. I would have to lie very still for a long time.

I am in charge of distributing the food for the chemotherapy group. Others cook it, but I serve it to the people. I decide to skip school the next session and just do the chemotherapy, more important than school to me.

The opening of the dream suggests that I am not ready for chemotherapy again. My body was still disoriented from the previous treatment and did not yet know how to get back to normal cell functioning, since my white blood cell count was still too low to be safe.

Underwear is the first thing that is put on when someone gets dressed, so the image of going off without my underwear (how embarrassing!) could show lack of emotional undergirding for the next treatment. Hair is often a symbol of spirituality, and the fact that my hair was not done in the dream might indicate that I was not spiritually prepared either.

The next paragraph of the dream could be viewed as a metaphor for chemotherapy. I am asked to be a human wheel, rolled along while others watch and film the process. Going through the chemotherapy felt as if I were rolled through physical abuse while others watched and wrote down everything they saw on my medical chart.

Since I did not think my body could withstand that harsh physical treatment, I asked for another part; however, the only alternative was the part of a dead person. The dream presents the two awful choices I felt I had: either a physical treatment that was too harsh for my body, or death.

The final paragraph showed me giving nourishment to other people going through chemotherapy, a picture of the kind of reaching out to others that sharing my story provides. The food I served in the dream was cooked by others, and the methods I learned from the healers I came in contact with were not invented by me, but were the fruit of the labor of many other people. My service in the process was to be in charge of distributing this nourishment to others going through cancer.

The dream closes with a statement that reflects my true values: surviving cancer is a higher priority than earning an advanced degree.

Handling the irritable mood mentioned by the dream was one of the most difficult parts of the chemotherapy package:

January 25, 1992, Journal:

I feel too scared of dying of cancer to quit the chemotherapy. Yet I feel at the moment that I can hardly bear the physical upset that the chemo does to my physical and emotional system.

I feel IRRITABLE. Every little thing gets to me. This morning the temperature in our house was 56°, and I got chilled. When Soli did not make the room hotter and told me not to fuss over the temperature, I got a severe attack of anger. I was thinking to myself, "I may catch pneumonia here in my own home!" I went to the sauna to get warm and wept while I was there.

When I returned from the sauna, the room temperature had dropped to 54°. I had a long talk with Soli, letting him know that I needed to be kept warm and that his frugality was the one thing I felt might break us apart. I could not live in the Spartan manner he liked, and I expressed that idea forcefully.

Directly confronting Soli was a new and awkward pattern of behavior for me, but I needed to give voice to my own needs. After we talked, I understood that Soli was trying to be true to his own ecological values by waiting until I returned from the sauna to fully stoke up the wood burning stove, not wanting to waste wood and contribute to global warming while I was gone anyway. He had tried to light a warm fire when he got up that morning and again while awaiting my return, but the wood was too damp to burn well, and he had not realized how important the fire was to me.

The following night, an affirmation of changing a pattern came in a dream in which I was giving advice to my cousin, whose time here on earth was about to be up:

January 26, 1992, Dream:

My cousin is quite centered on eating. I know that his time is up on March 8 or March 10. I speak to him about this transition and tell him that if he totally gives up his old life, he may be able to transform and continue his physical life in a totally new form.

The initiation was calling for a new pattern of behavior in me, totally giving up the old mindset of suffering in silence when my own needs were not being met. Soli was a good "sparring partner" with whom I could work on my issue of becoming more assertive since he loved me deeply and was committed to caring for me (and the world) in the best way he knew how. Once I spoke up forcefully about needing a higher room temperature, he kept the fire hotter for me.

As part of learning to love myself, I continued to open myself up to receiving help and healing from others who came into my life.

January 30, 1992, Journal:

I have felt significantly better ever since the cranial manipulation Marie gave me on Sunday, when she felt a blockage of something that made me cave into myself, made me be defended and shutting others out, from an early age, maybe even two or three years old. As I worked on the emotional issues that came up, Marie moved out the bones that had been scrunched up at my temples. Afterwards, my face felt wider, flatter, and more open.

Marie said that sometimes as she does an adjustment, she can feel the face beneath her hands changing shape to that of a child, and then she knows that she is working on deep childhood issues.

She said that the purpose of cranial manipulation was to clear the physical body so it could be filled by the spiritual. I felt deep inner relief after the treatment from her, a profound inner sense of relaxation and quiet. I could cope with my feelings better, and sexual energy also came back.

Another method of improving my health came into my life that January when I began an eight-week class in "Stress Reduction and Relaxation" at the University of Massachusetts Medical Center. This course greatly deepened my practice of meditation and yoga. In the stress reduction classes, Dr. Jon Kabat-Zinn taught that meditation was not only something to do sitting quietly for 20 minutes a day, but also something to incorporate into every aspect of daily life by becoming mindful of *being* in each moment of the present, which is the only time we ever have (Kabat-Zinn, 1990).

I realized that what I had formerly been calling meditation was actually visualization work in which I was consciously connecting with the Divine within me. In "mindfulness meditation" sitting practice, I learned the technique of quieting my mind by focusing on my breath and existing in a state of what Kabat-Zinn calls "choiceless awareness," not striving for anything, allowing the world to be exactly as it is.

The stress reduction tapes helped me keep on an even keel emotionally for the third chemotherapy treatment, which fell on January 31. I tricked myself into not worrying about it by telling myself it might have to be delayed another week because I had a head cold; however, the doctors proceeded with the chemo because my white blood cell count was up above 5. Physically, I felt worse from the sneezing and nasal congestion than from the chemo; emotionally, I felt serene from using the stress reduction tapes.

The hypersensitivity to cold kicked in again immediately after the treatment, so I stopped by the wig shop on the way home and bought two turbans to keep my head warmer.

February 5, 1992 (Wednesday), Journal:

The third chemotherapy treatment last Friday went quite well. I took the Zofran as directed and had almost no nausea at all the first two days. I smoked about 3/8" of a marijuana joint on Saturday night, which I believe helped cut the nausea on days three to five. I took Ativan and Compazine both on Saturday at 12:00 noon and 7:00 P.M. and on Sunday at 8:00 A.M. After that I couldn't make myself take any more Ativan, but I took an additional dose of Compazine at 6:30 P.M. on Sunday and on Monday at 12:30 P.M., as directed.

Tuesday I felt slight nausea and slightly more fatigue than usual, but the symptoms were quite tolerable.

During the past two weeks, I have been telling my clients about my having been through cancer, and I feel relief at not keeping the secret anymore. The clients have been kind and compassionate in response.

Tuesday night the acupuncture treatment got my system all tuned up again, and today I feel fine.

This last chemo was more difficult from the symptoms of the cold I had going into it than from the chemotherapy itself.

I continued to feel good between the third and fourth treatments, with almost normal physical energy, much joy, and no depression. The following dream added a bit of humor to my life:

February 16, 1992, Dream:

Vivien has two friends over, and they both want to date me. I feel complimented, then later inquire whether they want to go out with me as a sociological study of how old people are or whether they find me attractive just as I am.

I see sex as a sacred experience.

February 17, 1992, Journal:

An inner sense of completion fills my being. I FEEL healthy and happy. Great joy floods my soul. I experience myself in the flow of Divine Love.

The dream I wrote down the morning of February 18 did not make sense to me until I spoke with Suki that same night. The dream was as follows:

February 18, 1992, Dream:

Vivien is all dressed in her wedding finery. She looks beautiful, but I think the crystal pins she has on her cheeks with the pins going through her cheek flesh must hurt. She doesn't seem to mind though.

During the acupuncture treatment, I asked if points were ever put into the face since the chronic sinus congestion I had been having for a long time was not going away. I told her about the dream, and she put needles into my cheeks right on the spot where the pins had entered, medial side, on Vivien's face. She told me, "You dreamed your own diagnosis." I could physically feel the energy penetrating my sinus cavities during the time the needles were in my cheeks, and after the treatment, the condition began to gradually improve.

Only three weeks had elapsed since the third treatment, and on February 20 I gave a blood sample to see whether my counts were high enough to do the fourth treatment on Friday, February 21. That night I dreamed the following:

February 21, 1992, Dream:

A woman calls and says my lab tests came out great. Soli's friend calls up and says something about my having chemotherapy today.

Before I even called the doctor to find out the results, I was telling everybody that I thought my blood counts were high enough to get another treatment, and they were. The white blood cell count was at

3.6, just over the limit of 3.5 needed to go ahead.[1] I felt happy that I could get the fourth treatment over with, so I would be two-thirds done with chemotherapy.

The fourth chemotherapy session was the most difficult of the six rounds. The night after the treatment my mind was scrambled, and my dreams were jumbled, disorganized, and weird. I stumbled in the bathroom and fell onto the windowsill, hurting my lip. I could not focus enough to follow my guided meditation tape, and my third eye kept forming pictures that carried me away.

February 22, 1992, Journal:

I've felt yukky all day. The nausea creeps into everything. Just looking at the Ativan bottle increases the nausea, and as I took the last half tablet tonight, I nearly vomited from swallowing the pill. Then our puppy Terri came and sat on my abdomen to heal me up. I was touched by her sensitivity and love for me. She was biting at the fleas so bad, though, that I asked Soli to give her a flea bath. He did, and then I killed eleven fleas from the towel and one leftover from her tummy.

That night I talked on the phone with Vivien's paternal grandfather, and he told me the history of Vivien's grandmother getting cancer. She had found a walnut-sized lump in her right breast one morning, and the doctors sent her for a mammogram immediately. The results showed that the lump was a cancer. They operated very soon and did a mastectomy, not recommending radiation in her case since she was nearly 80 years old. The lymph nodes were all negative, and back then they did not recommend chemotherapy for node negative women.

---

[1]The white blood cell counts from the previous treatments had been as follows:

5.6 Treatment 1, (12/13)
4.0 Treatment 2, (1/3)
3.3 (skip) 1/24
5.6 Treatment 3, (1/31)
3.6 Treatment 4, (2/21)

Several years later she got bladder cancer, and the doctors said this tumor was not related to the breast cancer. But as Grandpa said, "Who really knows?"

The bladder tumor had already invaded the bladder wall and spread to the lymph nodes. She survived about one year from the diagnosis of bladder cancer, just about exactly what the doctors said she could expect to live. Hearing the details of the course of her illness made me grateful that I was getting chemotherapy, even though I was suffering from its acute side effects at the time.

The nausea continued, and by the third day after the treatment, I wound up vomiting water. Several factors may have contributed to making this round of chemo the most difficult one. First of all, before the chemo, I had eaten food that was too heavy: squid the night before the treatment, and a pancake the morning of the chemo.

Second, although the doctors had said to drink lots of fluids, the thought of drinking *anything* made me feel like throwing up, so I decided not to force myself to drink liquids and instead to respect the wishes of my body not to have much fluid intake. That decision was definitely a mistake! When in good health, the body will crave what it needs; however, my homeostasis had been knocked out of kilter by the drugs used, and I needed liquids to flush the poisons out of my system. The nausea produced by the drugs blocked the message that I needed to drink, and since I did not take in much liquid, the poisons stayed in my system longer.

Third, although the nurses had also told me to go home and take it easy right after the treatments, on the way home from this session, I stopped at my agency to move my things to a new office. Both the mild physical exertion from carrying boxes plus the emotional upheaval of moving my sacred office space somewhere else may have had a detrimental effect.

Fourth, I lost the herbs Suki had given me for nausea during the previous rounds of chemotherapy. And fifth, instead of having my pre-chemo acupuncture treatment the night before, as I usually did, so that my body would be in top form for the chemotherapy, I had the acupuncture three days before the chemo. After the difficulty I had with this session, I decided to wait four weeks each for the last two rounds of chemo to give my body more time to recuperate between treatments.

By the fifth day after chemo, I finally felt good again.

February 28, 1992, Journal:

Yesterday all day I was amazed at the synchronicities of the material my clients presented with what I had been reading the night before about temporal lobe epilepsy producing "hallucinations." And client images surfaced that exactly matched my dreams of the previous two nights.

What is going on? Am I more connected now than I was before?

I have felt really good, totally well again, since Wednesday.

Today when I was taking my fifteen kinds of supplements, I asked myself why I was doing all that, and I answered, "To take special care of my body during chemotherapy." I feel I am learning SO MUCH during this time, about myself, about my body, and about my connection to my Divine roots.

I feel no depression today.

The special day of February 29, which comes only once every four years, marked another milestone on my healing journey. Soli and I went to a surprise birthday party, and I got all dressed up and felt great about how I looked. In the conversation at the party, I did not mention cancer even once and did not feel the need to. We went as a normal couple and enjoyed the festivity.

Vivien told me of a dream she had, which reflected her confidence that I would survive the cancer:

Vivien's Dream, March 3, 1992:

I am with Mom for her last chemotherapy treatment. She is lying down on a bed like she did for radiation. They put fluid in her breasts and in her stomach somehow without needles.

She says, "Look, I only need this little bit for the insurance that the cancer will be killed." But then she does the rest just for extra insurance.

Encouragement continued to come from Vivien and Soli, from my inner life, and from the medical profession. During the all-day meditation session with the Stress Reduction program on March 7, 1992, I had the following sudden insight:

> "I realize who I am. Inside, I am a doctor, one who seeks to heal. I do not need to strive to become a doctor: the outer degree will be only a reflection of who I already truly am."

I felt delighted when Dr. Orr said that my breast was healing up wonderfully from the surgery. He optimistically added that my breast may continue to normalize over the next couple of years, but the more conservative oncologist said the left breast might retain its present swelling forever.

I decided that my breasts were like fraternal twins. They used to be identical twins, but then they individuated. Each one was beautiful in her own right. Since I had been applying castor oil to Leftie several times a week, the swelling in the nipple area had improved at times. She was slightly darker and firmer than Rightie, and the nipple was bigger and aimed downward, but she was still there, and she was beautiful!

Just as the doctors had warned might happen, my body went into crash menopause with the fourth round of chemo. "The (chemotherapy) drugs can create a chemically-induced menopause, with hormonal changes, hot flashes, emotional mood swings, and no period" (Love, 1990, p. 321). Before menopause, I used to get severely chilled and not be able to get warm; however, after menopause started, I got hot flashes during the day and frequently awakened at night feeling as if I were burning up, as if I were one of my pots inside of the initiation kiln being fired up to my melting point to make me stronger. I felt an insidious depression trying to creep in again.

As the chemical cycle progressed into the most dangerous part of chemotherapy, which is not the actual treatment, but the time in between rounds when immunity is low because of lack of production of white blood cells, a friend of Soli's called me up, encouraging me at a critical time when I was feeling bad. This friend was a cancer survivor who had been through eighteen months of chemotherapy. Out of compassion, she took the time to help me sort out the issues going on with me at the time. I listed in my journal the following points that our conversation helped me see:

1. Feeling shitty is part of the process of chemotherapy and is an important sign that the chemo is working. The blood has to get broken down to get rid of the cancer.

2. *Grace and Grit* (Wilber, 1991) is not the best thing for me to read at this point because it is the story of someone who died from breast cancer and is from the point of view of the support person. My focus at present needs to be on my own healing, not figuring out how to help Soli through the process. Soli needs to find his own support outside.

3. The body's metabolism is getting changed, so shifts in body chemistry and cravings are to be expected.

4. When I said I have to think of what if I wind up in the 10% of women who die from breast cancer, she nailed me on my "What If" thinking and said, "What if you get hit by a car?" The point is to live in the present moment as fully as possible.

5. She reminded me that nobody has told me I will die: nobody has indicated my case is terminal.

6. She also finished a degree while she was going through chemotherapy, and she understood my desire to finish the doctorate.

I heard myself saying to this friend, "I want my body back. I want to feel bouncy, but I don't have the energy."

Going into menopause also had an effect on my level of sexual desire:

March 13, 1992, Journal:

I experience my sexuality differently now because of all the changes in my life. Going through cancer treatment is a daily battle against the physical and emotional forces of illness and the side effects the drastic treatments have in my body. Killing only the bad cells and leaving all the good ones is impossible; I need much of my psychic energy to rebuild the damage done to healthy cells to be able to keep body and spirit together.

Intimacy is difficult at times, as I feel a strong need for private, quiet space in my life. Once in awhile desire for sex comes strong and hot, but usually something inside of me resists the intensity of the interpersonal contact. Often I do not want to reach orgasm during the process because it feels to me like losing something.

As I headed into the fifth chemotherapy treatment, I noticed subtle changes in my body from the medication I took the night before the treatment. After taking a whole milligram of Ativan, my dream recall was fuzzy, and I had difficulty thinking clearly.

March 20, 1992, Journal:

Melancholy is a good word to describe my feelings this morning. I fight with my feelings about having another chemo session today. I don't want all the side effects, especially nausea, but at a deep level I do want the medicine for the extra security it gives me of leading a cancer-free life.

Actually, I know the cancer is gone, but I just have to finish cleaning up the mess left behind.

My body is in good shape with the four-week interval. I have been dancing and doing aerobics this past week. My emotional state got balanced and centered with the cleansing by Tom, and I basically have felt my body return back to normal, except for menopause.

My son Robin came to visit for a week on his spring break to pull me through the agony of the next-to-the-last chemotherapy session. He arrived the day after my chemo treatment, when I was feeling pretty bad. I told him his job was to cheer me up.

By the time he left, I was feeling really good again. I especially loved hearing all his stories about his five months in China. He tried to teach us to sing "Happy Birthday" in Chinese, but I absolutely could not remember those strange sounding words. I was amazed that he learned to speak the language in such a short time since nothing in Chinese resembles anything in English except the one word "Mama."

By the third day after the chemo, the anti-nausea medications had me constipated.

March 23, 1992, Journal:

How do I unlock the grip of the iron fist around my belly? Yesterday I tried eating some marijuana. While it may have helped the nausea some yesterday, it also disturbed my sleep cycle. All night I dreamed of danger approaching.

All morning I was on the edge of vomiting. My abdomen feels obstructed. Some relief came after my bowels finally moved again.

But having Robin here with me is an existential moment. Today as I watched his nimble fingers playing Chopin, I marveled at what a fine young man he is.

Going cross-country skiing today with him was great fun. The three inches of snow here makes the countryside into a winter wonderland.

I noticed that in the midst of the nausea, as I talked with Robin about how GOOD I feel when I recover from the treatments, my body started to feel better immediately, just by connecting with the thought of being recovered. I was amazed at the positive side of the power of suggestion. On the negative side, I felt nausea just looking at the clinic where I got the treatments, but I could also move into a feeling of well-being by focusing on the feeling of the happy times after each chemo session is over.

As I sat by the lake waiting for Robin to get a different pair of ski poles, I consciously drew in Divine Love and let out nausea. I felt more energy coming in, yet I felt impatient with the nausea. I wanted it to go away and stay away forever, but I also felt thankful that the acute part of the treatment was nearly over and that only one more chemo remained. A week after this round of chemo, I went out to California again for doctoral classes and felt terrific.

Two weeks after the fifth chemo session I went to a wedding shower for the dear friend whose shoulder I had cried on the day I found out I needed to go through chemotherapy. At the shower, another breast cancer survivor friend who had just finished her radiation therapy three days earlier was telling people, "We're cured. No more cancer. We're just doing the treatments for extra security."

The other women there marveled to us, "You two don't look sick. You look wonderfully healthy!"

We told them, "We are!" and that our secret was "the Cancer Beauty Treatment!"

While we were delighting in our health, I felt a radiant healing glow in my entire body, and I was actually slightly amazed at how wonderful I felt. Having a normal energy level and an excellent mood were precious gifts, and I gave deep thanks for my healing, both spiritual, emotional, and physical.

My mood plunged again when the part of the cycle with lowered immunity set in, eighteen days after the chemo:

> *April 7, 1992, Journal:*
>
> I am tired, an existential tiredness that comes from my insides, from my very bones. Even as I sit at my computer writing this passage, my words are interrupted by tears.
>
> Today I KNOW again that the chemotherapy is working because I feel it breaking down my blood, breaking down my energy again.
>
> No wonder my sexual desire has been hypoactive. Not much juice is left in the blender!

As I felt myself going through a dark part of the initiation passage again, suddenly the full force of the synchronicities of the dates involved in my illness hit me:

August 14, just four months after our wedding date of April 14: cancer diagnosis

Friday the 13th of September: radiation tattoo

Friday the 13th of December: first chemo

Good Friday (Crucifixion day): last chemo coming up, April 17

I felt I was being crucified in my very bones, and out of that pain I wrote a song with the following words:

THIS MOMENT

How long will this journey be?
    I don't know;
    I really don't know.
How hard will this journey be?
    I don't know;
    How could I know?

Only the eyes of God can see
What lies in my destiny.
    I don't know;
    I don't need to know.

What will come in the years ahead?
    I don't know;
    I really don't know.
Down which path will I be led?
    I don't know;
    How could I know?

All I ever have is today;
I'm alive in the present moment:
    This I know;
    This I know.

So, come and join with me,
We will sing through eternity.

Come and dance with me,
We'll give thanks for the Gift of Life
    In this moment,
    This precious moment,

    Which is all we have,
    All we have.

    All we ever have.

The joy of the resurrection was already present with me on Good Friday, April 17, when I went in for the final chemotherapy session.

> *April 17, 1992, Journal:*
>
> Joy bursts forth from my soul. I am filled with the love of Lord Jesus Christ, grateful for my life and for my healing. I have no fear of this final chemotherepy session. Such happiness filled my soul today as I sang praises to God!

Going into the doctor's office, I felt wonderful, full of happiness that at last I was at the end of these treatments. My white blood cell count was up to 4.0 again, and the other numbers on the bloodwork were also good.[2] Then during the Cytoxan IV, I began to get nauseous, and I vomited in the car on the way home. I vomited again when I tried to take the second dose of Zofran, the anti-nausea medication; but although I vomited twice, I did not have as much nausea as in previous treatments. I did three things differently in this final treatment, which may have helped:

First, I ate enough dried fruit beforehand so that my bowels were quite loose going into the treatment, and I managed to eat four or five dried prunes each day, which kept my bowels moving once per day. Thus, I avoided the feeling I always had in previous rounds of chemo of having a rock in my belly. I still had some nausea, but not the rock effect. Since the Cytoxan is partly excreted through the bowels, keeping that system moving seemed to help alleviate the nausea.

Second, I baked a batch of oatmeal cookies and threw a half cup of old dried up marijuana into the batch. After chemo, I ate one small cookie no less than four hours apart and usually around eight to ten

---

[2]Specifics on my blood count from April 17 were as follows:

WBC = 4.0 (White Blood Cell Count)
RBC = 3.71 (Red Blood Cell Count)
HBG = 12.3 (Hemoglobin)
HCT = 37.2 (Hematocrit)
MCV = 100.2 (Mean Corpuscular Volume)
MCHC = 33.1 (Mean Corpuscular Hemoglobin Concentration)
PLAT = 192 (Platelets)

The RBC was slightly below the normal range of 3.9 to 5.2, and the MCV was slightly above the normal range of 81–99, but all of the other counts fell within the norms.

hours apart. I did not feel any "high" from the cookies, but they seemed to help settle my stomach. However, by the end of the third day, they tasted so raunchy to me that I could not finish them. The rest of the cookies wound up getting fed to my friend's dog, who got quite excited about them.

Third, with the approval of my doctor, I substituted Benadryl for Ativan (lorazepam). From my clinical experience, I knew that nausea was sometimes a side effect of taking Ativan. I had noticed a severe reaction of nausea in the previous treatment whenever I even handled the bottle of Ativan, but I did not have much trouble swallowing the larger Compazine tablet. As the treatments progressed, I had increasing difficulty preventing my throat from gagging on the tiny Ativan tablet, so I decided to do one of my research papers for school on the neurological pathways involved in the response of nausea to better understand the theory behind using Ativan, a minor tranquilizer, as an anti-nausea drug.

One of the primary reasons oncologists prescribe Ativan is to induce amnesia, to make the patient forget what had happened during chemo. Since nausea is a conditioned response, the body will activate the nausea pathways automatically with any stimuli associated with chemotherapy. Just driving by the oncology clinic on my way to work gave me a wave of nausea, even on the days after I had recovered from the treatments. Because the conditioned response builds over the course of treatment, by the end, oncology patients may be throwing up when they first walk into the office, before the IV is even in. The power of association is so strong with nausea that even just writing this chapter about chemotherapy produced mild sensations of nausea in my stomach, and I had to stop several times to push on the acupressure wrist points for nausea.[3] Forgetting the chemo lessens the buildup of nausea as a conditioned response.

Nausea was the only side effect of the chemotherapy that I really suffered from, and I was willing to take all three kinds of anti-nausea

---

[3]Suki had taught me to help relieve nausea by pushing on the center of the wrist, just inside the pulse point in a little pocket just under the longest crease line on the inside of the wrists. Acupressure wrist bands are now commercially available, marketed as relief for travel sickness and sea sickness. These bands are worn around the wrist with a bead pressing in on the acupressure nausea point and can be used during and after chemotherapy. They are also effective during pregnancy to relieve morning sickness. The bands could be hand made by attaching a small convex bead, about 1 cm in diameter, to a bracelet of elastic, with the bead pointing into the wrist.

drugs offered to me: Compazine, Ativan, and Zofran. However, these psychotropic medications also had the side effect of causing mental confusion. While under the influence of the combination of these drugs and the chemotherapeutic agents, my dreams were chaotic and contained frightful images of cruelty. The night after a chemotherapy treatment, I dreamed the following:

> I see my neighbor lady kicking one of her two young daughters. I tell her to stop and say, "You are putting me into a difficult position. I am a mandated reporter for child abuse." She says she used to be much worse and has improved a lot.
>
> They clarify my job as a social worker in their minds and realize the significance of mandated reporting. I say that since she realizes the consequences now of continuing to abuse her daughter, she will have reason to change.

I momentarily surfaced from the dream with many thoughts in my mind, like, "If you don't want your children, the state will find somebody who does," and, "Your children did not ask to be born. You brought them into the world, and they are your responsibility."

Then a deeper level of meaning set in, and I thought, "I wonder if the abused little girls in the dream are parts of me—that chemotherapy kicks me around, and my inner child aches."

A child thinks and lives in the present and cannot understand things like long-term gain from a harsh medical treatment. The adult in me knew that undergoing chemotherapy was an act of love for myself, but the child felt as if she were being battered.

I realized that I had improved a lot in the care of my inner child, just like the woman in the dream was better than she used to be; however, I was "mandated" to completely clean up the self-abuse. I felt that if I did not learn to treat myself right, the cancer would be more likely to return, a realization that motivated me to improve my behavior towards myself.

> *Further dream:*
>
> I see my inner male in bed crying. I go near him, and he confesses to me, "I can't think!" He has been plagued by this disorder for quite some time. Some other clinicians try to decide if it is related to his drinking diet Pepsi, a caffeine-induced organic problem.

Although in real life I do not use caffeine, the dream points to an organic cause of the mental difficulty, as caffeine is another mind-altering drug. I asked myself why I was taking Ativan, an anti-anxiety medication, when I did not have anxiety about going through chemotherapy. I wanted the treatment; I believed it would help me, and I was at peace with getting it. I wondered what anti-anxiety medication would do inside my body if it could not find any anxiety to alleviate.

Even though Ativan theoretically helps prevent buildup of the conditioned response of nausea by causing mental confusion and making the person forget what happened while under its influence, since Ativan can also cause nausea, I decided to skip taking Ativan with the final treatment and replace it with Benadryl instead, and the nurses agreed to this substitution.

After the treatments were all completed, I realized that I hated periodic vomiting less than I hated constant nausea for five days. Without the Zofran in the first treatment, I had episodes of vomiting for the first two days; however, the nausea came only right before the vomiting and then was over with after I threw up. Also, the nausea and vomiting were over with after two days in the initial treatment without the Zofran, instead of dragging out low-grade nausea for four or five whole days as in the rest of the treatments, which included the expensive Zofran.[4]

My extended family had a gathering Easter Sunday, two days after the last chemo, to celebrate my resurrection from cancer. My body felt lousy the whole day, but I enjoyed being with the large circle of family and friends that gathered to honor the process of new life that follows death, and I was overjoyed that I did not have to go for any more bear-banging, child abusing sessions of chemotherapy.

Four days after the last chemotherapy, on April 21, I stopped taking the anti-nausea medications and went for the acupuncture treatment documented in Kirlian photographs 17 and 18, where I talked with Suki during the treatment instead of meditating.

Suki concentrated on bringing up the energy in my damaged kidney and liver meridians in the acupuncture treatment, and after the

---

[4]Note: Each person's response to a drug will be different, so I am not making any recommendations for or against specific medications. Another person might not have any nausea at all with the same combination of drugs that I used, but the doctors and I had to learn how my body was responding to the drugs. The patient and the physician need to work together to monitor the effects of the medications and adjust them for the best possible results.

combination of acupuncture in the morning and a full-body massage in the afternoon, I felt as if I had arisen from the dead and were in heaven!

April 21, 1992, Journal:

The weather today was the warmest it has been so far this year. Several of my new daffodils I planted the first weekend after I started radiation are about to bloom. The rhubarb is up, and the dandelion greens are starting.

But the tree leaves are stubbornly refusing to come out. What are they waiting for? Do they want to delay appearing till they are sure I am all through chemotherapy? I will shortly be ready to begin a new life.

I was touched by the note a friend who is also going through breast cancer sent me and felt surprised by her talking about my remarkable attitude through this whole ordeal. I think anyone else in my shoes might do just exactly as I have done.

NOW I AM READY TO BEGIN REGENERATION! I got back to meditating again today. I will care for my body lovingly, to get that abused child I saw in my dream safe and secure and filled up with love. *I want her!*

CHAPTER 17

Losing My Hair

Although the nausea from chemotherapy had been difficult, it was also very temporary; the only side effect of chemo I really feared and dreaded was hair loss. I did not want to be bald! When Dr. Orr had recommended I get chemotherapy, my first question to him had been whether I would lose my hair, and my heart fell when he said I probably would.

Cancer cells divide more rapidly than most normal cells, so the way chemotherapy works is that the chemotherapeutic drugs kill any cell that is in the process of dividing, by blocking the metabolic process of the cell. Chemotherapeutic drugs are non-specific and attack any dividing cell, not just cancerous ones, so other areas of the body where cell division is rapid are hit hard: the lining of the stomach; the bone marrow, which produces white blood cells; and the hair roots under the scalp, which are constantly producing new cells at the base of the hair follicle to push the old ones down the strand of hair.

Several kinds of chemotherapeutic drugs are given together to attack cells in different stages of division. The combination given me was the most common recipe used for adjuvant chemotherapy: Cytoxan, methotrexate, and 5 fluorouracil (CMF). Different drugs produce varying amounts of alopecia (hair loss), with the most hair loss coming from the stronger drug Adriamycin (doxorubicin) substituted for methotrexate (CAF) in late stage cancer where tumors have already been found in other parts of the body. Adriamycin always produces hair loss, but methotrexate by itself only rarely causes hair loss. The alopecia in the CMF combination comes from the Cytoxan and 5 fluorouracil, which will often produce hair loss.

Although only 15% of the women receiving the milder CMF combination lose all of their hair, the hair almost always thins out, often to the point of needing a wig until the hair grows back after the treatments have finished. Knowing these facts, Dr. Orr and the oncologist wanted me to be prepared for hair loss.

The first step I took to come to terms with the fear of my hair falling out was to directly face this fear in meditation. As I got quiet inside, I remembered from my studies at the Jung Institute that in ancient Greek society, before a young woman could marry, she had to be initiated into womanhood by sitting on the steps of the temple of Aphrodite and having sex with the first man who threw a coin into her lap. After that experience of sacrificing her virginity to the goddess, she was considered invulnerable to being seduced. The alternative to serving as this "sacred prostitute" was to cut off all her hair. A form of this ancient ritual has continued into modern times in the practice of the Catholic church requiring a woman to cut off all her hair as part of the initiation into becoming a nun, sacrificing her hair to God.

I made a conscious decision to give up my hair as a sacrifice to the goddess for my initiation into a fuller sense of womanhood. To soften the blow of the loss, I asked God, "If at all possible, could my new hair that grows back please be naturally curly, like a friend of mine got after her chemotherapy?"

Once I had confronted and released the fear of baldness, methods of keeping my hair came to my attention. In the attunement in October, Gloria advised me to begin praying daily, "Thank you, God, that my hair is not lost. Thank you for my strong and healthy hair roots." Every day in meditation, I visualized the cells in my hair roots being vibrant, full of life energy, and making strong hair. For further protection, I had my hair cut short before chemotherapy, as short hair is less likely to break off at the root, where the chemotherapy attacks it. I also had my hair trimmed every month on the full moon, as folk wisdom in many cultures says that hair gets stronger if cut on the days around the full moon or the new moon.

I still kept the wig the oncology department gave me, since the doctors all agreed I would need it, but my heart resonated with the daily visualizations of healthy hair and my friend Starr's positive affirmations about keeping my hair.

I considered using an ice pack on my scalp during the chemo sessions to slow down the circulation of chemotherapeutic agents through my scalp area but decided against that plan because the nurses told me that if a cancer cell happened to be in the scalp area, with the ice pack on, the chemo might not be able to get in to destroy it.

When I went to see my oncologist on January 25, before the third chemotherapy treatment, I proudly informed him that none of my hair had fallen out. He said, "I still think you will lose your hair." I asked when he thought it might come out, and he answered, "It can happen any time." At first I felt sad and lost my image of keeping my hair, but then I got so angry at his conservative, realistic response that I went home and prayed and visualized my hair even stronger, just to spite his negative attitude.

However, hearing what happens in most women let some doubt and fear creep back in, and the following dream I had the week after seeing the oncologist mirrored those feelings:

February 4, 1992, Dream:

As I wash my hair, big clumps of hair one to two inches long come out into my hands. I look in the mirror and see my hair has thinned way out and has bald patches on the top.

I awoke in horror and could not get back to sleep for a long time. This description matched the way the books had said the hair would probably come out, while washing or combing it. The image in the mirror frightened me. When I finally fell asleep again, I had the following dream:

I am in the office of Isabel, who is with another patient. *(In real life, Isabel was my ballet teacher, but in the dream she was a health practitioner.)*

Isabel steps on a machine, and I see colors light up all over her body showing the different meridians and areas on her foot like the diagram I have of foot reflexology. The client's areas do not light up, only Isabel's because she is in such good physical shape.

I ask Isabel if I have an appointment with her today and she says yes, at 7:30 P.M. I ask if she can see me now since I am there and have acupuncture tonight at 7:00.

Some other patients have been missing, so she has time for me. She asks what is on my mind right now, and I tell her about the dream of losing my hair. She says she will work on my head and begins adjusting it.

She asks the doctor named Ben to sign a prescription for three things for me: one is a long name I do not recognize, the second is a hair wash product by the name of PRAISE, and the third is a dairy product called something like a cross between ricotta and linguine.

After the treatment I have to fight to get into the shower to get ready for work.

The initial dream was a vivid picture of my worst fear, and the image horrified me so badly that I awoke from the dream because of the emotional impact of seeing my hair falling out. Earlier, I had faced and released my fear of baldness as an abstract concept, but this image brought me face to face with a deeper level of the fear, which included the emotional shock of seeing myself half bald.

When I finally got back to sleep, the dream that followed helped me work through the fear by showing me ways to avert the possibility of baldness. The first image the later dream presented was going for bodywork. Suki had been talking about the meridians of energy that flow through the body, but the idea stayed intellectual until the dream showed me the meridians and reflexology points in vibrant, living color. Then "the dream Isabel" adjusted my head, my cranium. She seemed to be a composite of many ways to work on the body: ballet dancing, acupuncture, reflexology, craniosacral manipulation, and conventional Western medicine.

Hair is located on the kidney meridian in acupuncture, and this meridian was usually in bad shape right after each chemotherapy treatment since excreting the chemotherapeutic drugs from the body taxes the kidneys heavily. Perhaps strengthening the kidney meridian with acupuncture also strengthened my hair.

The first item the doctor in the dream prescribed was likely a conventional Western medication, which usually has a long name I do not understand, and modern medicine played the major role in my overall healing, keeping my soul and my hairy body together with surgery, radiation, and chemotherapy. Without the drastic interventions of Western medicine, the cancer might have continued to spread; and without a body, I could not keep my hair.

The second item prescribed by the doctor in the dream is a shampoo named PRAISE: I was to wash my hair in praise, exactly what I had been doing with the visualization of giving thanks and praise to God for my strong and healthy hair. Perhaps an overall attitude of thankfulness and praise also contributes to strong hair; flowing with the moment and giving praise for whatever comes in one's life is a strong spiritual practice, and hair is used in the Bible as a symbol of spiritual strength and a connection to the life-bringing forces which flow from the unconscious part of the mind.[1] Praise strengthens spirituality.

The third item from "Doctor Ben" hinted at the need for some high quality protein in my diet. I had quit eating red meat many years earlier, and I had reduced my dairy intake drastically because of its high fat and high estrogen content;[2] thus, my vegetarian diet also avoided most high quality protein foods. I had been reading about how excessive protein in the diet can have detrimental effects in the body like inducing cravings (Colbin, 1986), increasing osteoporosis (Love, 1990), and increasing the overall risk of getting cancer (Robbins, 1987), and perhaps I was erring on the side of getting too little protein.

The dream mentioned "something like a cross between ricotta and linguine," and combining a small amount of a dairy product like ricotta cheese with a larger amount of a grain product like linguine pasta increases the quality of protein available for the body to use.[3]

---

[1]The story of Samson losing his strength when his hair was cut and regaining his power when it grew back is one example. The Biblical admonition for women not to cut their hair refers symbolically to their not cutting off their spirituality or their connection to the unconscious.

[2]Fat, estrogen, and breast cancer are all related. Menopausal symptoms such as hot flashes are caused by estrogen withdrawal. "Heavy women have fewer menopausal symptoms than thinner women, probably because fat makes estrogen and gives them a boost" (Love, 1990, p. 19). Fat cells make estrogen, and high levels of estrogen promote higher rates of breast cancer (see p. 57 and footnote p. 10). The higher the percentage of fat in a woman's diet, especially animal fat, the greater her risk of getting breast cancer. Eating meat daily quadruples the risk of breast cancer over eating little or no meat. And increasing consumption of eggs, butter, and cheese also increases the risk of getting breast cancer (Robbins, 1987). "Diets high in meats, dairy products and eggs not only force an early menarche, they also delay menopause" (Robbins, 1987, p. 266). And both early menarche and delayed menopause increase the risk of breast cancer.

[3]Frances Moore Lappé (1975) wrote extensively about combining foods to give an overall higher level of usable protein to the body without relying as extensively on meat and dairy products. The body cannot make eight of the twenty-two amino acids essential for protein metabolism, so these eight "essential amino acids" must come from the food we eat (p. 66).

Since hair is made of protein, perhaps the dream was suggesting that I needed to combine foods, possibly including a few dairy products, to ensure an adequate supply of protein to keep my hair healthy while I was rebuilding tissue from the cells killed by the chemo. In the long run, my need for protein would be lower after the chemotherapy was finished, and I could go back to a more strictly vegetarian diet to improve my overall chances of survival.

The ending of the dream commented on my struggle to maintain my work schedule during all of these treatments. As the dream says, I had to fight to keep going to my job, even though I felt connected to the team and the clients there.

My psychiatrist supervisor at work kept affectionately tugging at little pieces of my hair every month and marveling that it was still so firmly attached. I joked that probably the reason why my hair had not fallen out was because of the superglue I put into my shampoo. From his extensive experience in oncology, he said usually the hair fell out at the end of the six months of chemotherapy, when the chemical build-up was at its peak.

When I returned to the oncologist for the fifth chemo and still had not lost any of my hair, I expected him to once again predict it would fall out. However, to my surprise, his attitude had changed:

March 23, 1992, Journal:

The oncologist was more human this time during our appointment. When I said I hadn't lost any hair yet, he said, "Good for you!" And when I said that even if it did come out, it would grow back soon, he added, "It would be back by summer."

Our session was interrupted by a phone call from somebody whom he put on instantaneous morphine or something like that. He said over the phone that they had

---

Dairy products, such as ricotta, have a surplus of the essential amino acid lysine, and grain products, such as linguine, have a deficiency of lysine, so eating them together improves the overall amount of protein usable from the grain. However, the grain could also have been combined with a legume such as dried beans or lentils because legumes also contain a surplus of lysine. Dairy products are not necessary to get a high quality usable protein.

increased the pain medications, but that had obstructed the person.

I said to him, "That was somebody in worse shape than me," and he replied that one has no idea of some of the tragedy they get involved with at times.

A couple of weeks after the last chemotherapy treatment, I still had not lost any hair from my head; however, my pubic hair had begun thinning out slightly after the second chemotherapy treatment, and after the last treatment was finished, it thinned out even more. With a sudden flash of insight, I realized that when I had done the visualization of my hair roots being strong and healthy, I had been imaging only my head, not my crotch. I had used the PRAISE shampoo only on my head area.

When I had the radiation treatment on my left breast, the one hair I used to have on the left nipple fell out. In sympathy, the one hair on the right nipple also fell out during chemotherapy. Six weeks after I finished the chemotherapy, the one hair I had lost on the right nipple began to grow back, and the pubic hair also filled back in. But the left nipple has remained permanently bald, and the one hair I used to have there is the only hair I really lost. In the title of this chapter, "Losing My Hair," the word "hair" is singular.

I wondered whether my visualization of a head of strong, healthy hair could have prevented loss of hair from my head, since the unvisualized hair in my crotch thinned out—whether a positive thought pattern could have affected the condition of my hair.

I knew from placebo studies I had read that a placebo often got dramatic results, and the placebo was working on the power of thought alone. Norman Cousins (1979), a major pioneer in the field of psychoneuroimmunology, wrote, "The placebo is the doctor who resides within" (p. 69). Cousins cured himself of a crippling disease by tapping into the power of the mind to cure the body. He took charge of his treatment, enlisting his doctor's help to try new ideas, like using massive doses of vitamin C and moving out of the hospital into a hotel room. Cousins literally laughed himself well by watching old Marx Brothers movies. In his book *Head First: The Biology of Hope and the Healing Power of the Human Spirit* (1991), he relates the role of positive emotions in creating health and helping people to cope with and overcome illness.

I had also read studies by two mind-body authors, which reported that a placebo had *produced* hair loss. Bernie Siegel (1986) reported a study in England in which men were told they were given chemotherapy but instead received a saline placebo: 30% of these men lost their hair. And psychoneuroimmunologist Joan Borysenko (1988) reported a clinical test in which a third of the women who received a placebo instead of chemotherapy still lost their hair, both examples of the power of the placebo effect used to make a negative change. I started wondering how a placebo could make hair fall out and did some studying on how hair actually grows.

I found that at the base of each individual hair follicle is the papilla, which contains the veins, arteries, and nerves needed for hair growth. An artery below the follicle carries blood to nourish the papilla, which produces new cells that then push the old dead cells up the hair shaft. Each hair on the head grows at an average rate of 13 millimeters (one-half inch) per month for two to six years; *the hair falls out when it stops growing,* and then the papilla activates and makes a new hair. Many factors influence the activity of the papilla, "including age, diet, general health, and the condition of the skin" (*The World Book Encyclopedia,* 1977, s.v. "hair"). Hair needs to keep growing at its base to stay in the scalp, and its growth is influenced by the blood supply sent to it through the artery that goes into the papilla.

Blood supply is an autonomic function that is influenced by the emotional state a person is in. Fear generates a stress reaction in the body, and a stress state raises blood pressure and directs the blood supply to the muscles needed for fight or flight and to the muscles attached to the hair root, which make the hair "stand on end," not to the papilla for growing more hair.

I wondered whether the third of the placebo group who lost their hair might have had so much fear of baldness that this fear constricted the blood supply to their hair roots, closing down the nutrients to the hair follicles and making the hair fall out. The placebo is just a thought, and if the thought of hair loss could constrict the blood vessels to the scalp, perhaps the visualization of strong, healthy hair could open up the blood vessels to the scalp, nourishing the hair roots to make them stronger to counteract the cell damage done from the chemotherapy.

Perhaps fear gave the placebo chemotherapy its power. Fear tends to paralyze the person who has it from taking action to prevent the very thing that is feared, and fear also acts as a magnet, drawing into one's life the very thing one fears because of the powerful visualization of the feared event. (With this logic, one should fear riches, health, long life, and God!)

I had released the fear of baldness by facing the possibility; I prepared emotionally initially by being willing to sacrifice my hair and then by doing "praise" visualization, and I prepared physically by having a wig and some turbans on hand, cutting the hair short, getting the hair trimmed on the full moons, getting bodywork, and eating well.

I will never know for sure whether the visualization I used was the reason I kept the hair on my head when my pubic hair thinned out as expected; however, the fact that a placebo can produce baldness attests to the power the mind has to make changes happen in the body.

Stephanie Simonton, one of the pioneer researchers approaching healing from cancer holistically, is now doing a research project with people who have a rare form of cancer that starts in the salivary glands, for which modern medicine has no treatment. The core tool of her healing program is visualization (Simonton, 1993) because visualization is so powerful. The body seems to react to a picture of an event as if the event were really happening. It takes the placebo effect one step further, putting the healing into the mind even without the sugar pill.

C H A P T E R 18

Baby Dreams: New Life

Going through the initiation of cancer opened up new life within me; old patterns based on fear died and were shed, like a snakeskin that is split open and left behind when the snake grows. Leaving behind old, confining emotional patterns made room in my heart for a new relationship with *the Self*, the indescribable, numinous archetype[1] of wholeness,[2] and the divine child within.

Leaving behind my old skin (who my ego had thought I was) did not come easily. But in order to get through the initiation and birth the Self, to get myself centered in the fullness that Life was calling me into through my illness and my dreams, my ego needed to surrender control of the reins to my Dreammaker, the unconscious part of my mind, which was working hard to bring me into wholeness.

As I burned through the fire of radiation, swam through my own tears from its emotional ups and downs, and faced the life and death struggle cancer presented, I gradually released clinging to my orientation around my ego and surrendered control to the Self as "the new centre of personality which replaces the former ego" (Jung, 1970, p. 494). Although the story of this chapter is personal, it describes a process that is universal.

---

[1]Webster defines "archetype" as "the original pattern or model of which all things of the same type are representations or copies: PROTOTYPE." Jung frequently mentioned archetypes as being the archaic primal symbols or forces behind universal modes of behavior that are found across many cultures.

[2]Jung defined "the Self" as the combination of the conscious and the unconscious parts of the mind (1970), so its scope is much broader than just the individual ego, which is the central reference point of the conscious part of the mind. The Self is paradoxical in nature, being by definition "beyond the bounds of knowledge" (Jung, 1970, p. 63), since it includes an *unconscious* part.

The Self will not fit into any one container since it is a dynamic, moving energy. Therefore, many symbols are used to illuminate different facets of this profound archetype. Jung found that a circular or square mandala was frequently used to symbolize the Self because it represented the union of the opposites, the coming together of east and west, male and female, young and old, the divine and the human into a unified whole.

Alchemy, a medieval science which Jung studied, often represented the Self in cosmic symbolism using pairs of opposites, since the Self unites the conflicting opposites of the conscious and the unconscious parts of the mind.

> The intensity of the conflict is expressed in symbols like fire and water, height and depth, life and death. (Jung, 1970, p. 6)

My dreams chose another symbol of the Self, that of *the divine child* (Jung, 1970), to bring to my awareness the magnitude and importance of the process that was at work in my life.

For a pregnancy to begin, the opposites of male and female need to unite, and from this union, a new being is created. Years before I got cancer, I kept dreaming that I was pregnant, and as a woman who has experienced the miracle of pregnancy, birth, and breast-feeding of two real, live human children, this image really caught my attention. My dreams were telling me that a new energy was entering my life; a profound force had been planted within me and needed my care and nurture.

The energy of the divine child is a miraculous union of the human with the divine, a coming together of the conscious and the unconscious.[3]

My getting pregnant in the dreams was clearly a miracle, since in real life, I had a tubal ligation shortly after the birth of my second child. But my dreams were not daunted by the medical improbability

[3]Every story of the birth of a divine child contains some miraculous element, and often the father of the child is a god, while the mother is a human. In Greek mythology, Zeus fathered many children with mortal women; and Mary, the mother of Jesus, was a virgin who was impregnated by God (Luke 1:31–35). Mary's relative Elizabeth was barren and already old (post-menopausal) when the special child John the Baptist was conceived (Luke 1:36). And likewise, in the Old Testament, Sarah was already 90, post-menopausal, and Abraham was 100 when the angel of the Lord told them Sarah would bear the son God had promised to give to Abraham to make a great nation of him. Sarah laughed when she overheard the Lord telling Abraham she would get pregnant, and the name God told them to give to this special child was Isaac, which means, "he laughs" (Genesis 17:19 and 18:1–15).

of this image they kept giving me: the process of healing from cancer was birthing new psychological life within my soul. Ironically, one of the tricks that cancer plays to hide from the immune system is that it mimics the chemistry of a fetus (Quillin, 1994) to hide its abnormally rapid cell division. The fetus is the only naturally occuring normal development in the body where growth is as rapid as it is with certain malignant tumors.

This chapter traces the thread of images of pregnancy and birth woven into my dream life through a four-year time period, starting before the diagnosis of cancer and extending through the cancer treatment into the time of reorganizing my life after chemotherapy was over.

February 7, 1989, Dream:
(I was in the process of applying to the doctoral program but had not started it yet.)

I am in graduate school and am pregnant. I feel my belly swell with the new life and need clothes that will stretch. I will be glad when I've graduated from school so I can have a more normal life. I am in class, but feel I'm not concentrating on my studies. Then I think I am sure I am *not* pregnant.

Am I?

The dream showed me being pregnant and in school at the same time, a frequent theme in the dreams that followed, but the pregnancy was not the school itself; in the dream, getting the Ph.D. will be nice because then I can relax into a more normal life, but it is not the baby. I also needed clothes that would stretch because of the new life that was growing inside of me, suggesting that my self-image was too confining and needed to stretch out.

The conception itself was still a bit shaky in this first dream, because I was not concentrating on the right area of study yet, and I was unsure whether I was really even pregnant.

The ending of the dream hinted at the need for a firmer sense of self-awareness. In the Old Testament, one of the names God called himself was "I AM THAT I AM" (God is very sure of himself), and a present day religious group calls God the "I AM" presence. But in the dream I have things backwards, with the question, "Am I?" I was not sure whether I was pregnant, and in a larger sense, was not grounded in a firm enough sense of my own existence.

When I actually started the doctoral program in the fall of 1989, I had the following dream:

September 28, 1989, Dream:

I AM CRYING MY EYES OUT. I AM PREGNANT, NO FATHER AROUND. NO FAMILY SENSE.

I was in the position of Mary, pregnant without a father for the baby, except that the angel Gabriel had not come to tell me what was going on, and I felt lost. The new life was already planted within me, but the baby had no father around to form a firm grounding for the baby to grow. I felt intense sorrow at not having a sense of family support, not having enough strength in the male side of my personality.

I had not met Soli yet at the time of the dream, but the stage was beginning to be set for him to come into my life, like a Joseph who would help Mary care for and nurture the divine child.

I met my psychic friend Sara that month, and she suggested that the reason the relationships with men in my life never lasted was because I myself was resisting relationships; she encouraged me to go inside to find out what that resistance was all about.

In meditation, the insight came to me that my true path is a spiritual one, but because I had been so hungry for a man's love to fill up the part of me that felt unlovable, I had always let trying to please the man in my life sidetrack me from my spiritual center. I made a commitment to the Self within and vowed that if a man were to come into my life, I would keep the God within first and the man second. I also fully accepted the possibility that I might never remarry and fully accepted the difficulties and joys of singleness.

These new attitudes were reflected in the following dream, which shows union within:

October 11, 1989, Dream:

My inner husband and I are getting along well together, taking joy in the relationship. I am ready to share a baby together if he is.

I am sewing an orange skirt to wear to work and want to sew in a special pocket for dreams to put my dream notes in, to think on during the day.

The commitment I made to God was the real fathering of the baby, and this union with God freed up the energy for a human man to walk into my life. When I needed a man to love me because I did not love myself enough, the human job description was too big; however, once the internal connection with divine love was in place, then a man only needed to give the normal amount of human love necessary for a relationship to work.

The dream also connects inner union with working with dreams, incubating them in my pocket like a baby kangaroo in the pouch. Dreamwork is one of the quickest avenues to bring the content of the unconscious into the awareness of the conscious mind, giving a chance for the opposites of conscious and unconscious to integrate into a fuller personality.

The week before I met Soli, I had the following dream:

October 25, 1989, Dream:

I am almost full-term pregnant and think I am going into labor. Exciting! My husband says the main reason he is there with me is because of the hope of children together in the second marriage.

I think about being tied down again for another 18 years. But I am very happy too.

Later I look and am only about two months pregnant.

Another clue that this pregnancy is a divine child instead of a real pregnancy is the tricky quality of time, first being almost full-term and then being only two months pregnant. With a real baby, one cannot go from nine months pregnant down to two months pregnant!

Care of the Self takes time and is a long-term commitment of intensive care, just like raising human children. And just as the commitment to one's children lasts a lifetime, caring for the Self is also a lifetime task—but a joyful one, reflected by my happiness in the dream.

When I met Soli at a folk dance on November 3, 1989, he arrived just in time to lay the foundation of solid family support for the process that wanted to unfold in my life. I felt much joy at the connection with Soli and had the feeling that we had known each other for a long time already.

In January of 1990, when I went to visit the home of a client of mine who was crippled from a bout with bone cancer, she was very excited and told me she had a "revelation" for me. I felt very surprised since nothing of this sort had ever happened with any of my clients before; however, I felt especially close to this client because I had been her therapist during the time she had been diagnosed with bone cancer and had stood by her as she battled through amputation of part of her leg and chemotherapy. The transformation and deepening of the personality that came in this woman's life as she went through the initiation of cancer had deeply impressed me, and I had great respect for her. Without my having mentioned a word about a new man in my life, my client told me she had just dreamed,

January 15, 1990, Client's Dream:

I see Barbara with a man who is handsome, fair, tall, very nice to her, at a place like a hotel. He treats her very well, and has money, lots of money. He is also very nice to me. My body is whole and radiant, and I am serving Barbara and this man and being appreciated. There is much happiness for Barbara with this man.

My client added that sometimes, when people have a failure in a marriage, they might be afraid to try again; but she saw much joy for me in this situation and encouraged me to go with this man. I talked with her about how the dream also applied to her, reflecting a wholeness that grew within her as she had healed from cancer and had gotten together the masculine and feminine parts of herself. She agreed with me but insisted that this dream was a special message that God revealed to her to deliver straight to me.

The synchronicity of this "revelation" dream helped me confront my doubts about making a commitment again. In the deepest part of my heart, I still ached from the cumulative pain I was carrying from childhood wounds, and I was afraid that this new relationship would not last.

A vision of wholeness had been conceived within, the divine child, but this energy stood in stark contrast to the human child of my memory, the little girl within me who had been molested and whom I was neglecting because I was so busy focusing on the needs of everybody else in my life. Recurrent images of crying, abused children came in dream after dream, partly because I saw so much heartbreaking child abuse in my clinical work, and partly because I needed to become more loving to the child within me.

January 19, 1990, Dream:

My child is crying and crying. I go to comfort her.

The human inner child was crying because I had not learned how to truly love and care for her, how to put her needs on an equal plane with the needs of the others in my life.

January 20, 1990, Dream:

I see a trace of blood on my underwear and know for certain I am pregnant with Soli's child. I am talking with my client and tell her I had a tubal ligation. I'm not sure her daughters know what a tubal ligation is.

I am sorting through a bunch of pairs of pants; some would fit me for a short time only, and some have stretch panels to go for the whole pregnancy.

The miraculous element of conception in spite of blocked Fallopian tubes is reinforced in this dream, and the earlier theme of needing to stretch out to make room for this new life is also repeated.

March 9, 1990, Dream:

I am pretty sure I am pregnant. I think about how much work it will be being in the doctoral program and having the baby, but I want to keep going to school.

Caring for children, including divine children, requires work, and being in graduate school also required an output of concentrated energy. In the dream, I was struggling with the eternal dilemma of balancing my inner and outer needs. When I began the doctoral program, I did not cut back on my hours at the clinic where I worked; however, that year my daughter Vivien wanted to move to a distant state to spend her freshman year of high school living with her father and her older brother Robin, who was going into his senior year of high school and would leave for college the following year.

I agreed with Vivien's idea, feeling she needed time with her father too and that I would have more energy free to get adjusted to the demands of the doctoral program. But at the end of that first year, when Vivien informed me in June of 1990 that she wanted to stay with her father permanently, I felt crushed. She was my youngest child, my baby, and I was not ready to give up my mother job yet. I wanted her back!

I fell into a depression, unable to accept this loss and grieving over the time I felt I would be missing with Vivien. Not even the comfort of Soli's presence could take away the pain of the emptiness I felt.

August 8, 1990, Journal:

FAREWELL TO MOTHERHOOD

Children are the most precious gift God gives us.
But it's something I cannot hold in my hand.
I have to give my daughter back to the
Universe, to her father, to
Mother Earth.

Inside I hurt SO, SO BAD.

I want to keep her.
She is so beautiful.

She is not mine—
She belongs to God and to herself.

God, thank you for the very short fifteen years that I have been her primary caretaker.

I remember the day she was born. I hadn't quite finished harvesting my garden, and she came to grace our lives.

How I love that child!
How I grieve that she is leaving me.

I want to be a mother. It's been the only or almost only really wonderful thing in my life.

When I started, making room for Robin and Vivien in my life was SO HARD.

When I divorced from their father, caring for them was hard. How I ever found the courage to separate from their father, I still don't know.

Breaking apart those children's parents broke my heart. I remember how sad Robin was that next year and the psychic hell I went through.

I know so much about psychology, yet *I don't know why this hurts so much*. Having them was so hard—so very difficult for me to learn how to be a mother. And now that I've finally caught on and learned how, the job I love is being taken away.

It hurts right through, piercing my soul.
God, how I grieve this LOSS.

The agony of this moment of separation from my flesh and blood daughter was part of the dying inherent in the initiation process. The old ways and forms of life had to die to make room for the new relationship with the Self. As my inner child was incubating, my outer physical child was leaving me.

The night after I wrote this farewell to my mother job, I dreamed of an old woman, a post-menopausal crone, having a baby, showing that after the time of physical birth and nurture is past, birthing continues in other realms:

August 9, 1990, Dream:

My 75-year-old client who has a speech dysfunction has a baby. She calls the psychiatrist to come hold the head as it comes out, and my client speaks perfectly clearly through the whole thing.

She is right—the baby is ready to come, and she pushes it completely out. His face is simultaneously newborn and a wise old man. She cuddles him with motherly love and says, "Hello Baby." After a bit her husband comes over to appreciate his son.

This client was the very least likely person I could imagine to give birth, but in the dream, she miraculously has a child who is simultaneously newborn and a wise old man, the union of the opposites of young and old. The dream was telling me that as I myself made the transition from the mother stage of womanhood into the wise woman phase, new life would come to me.

Menopause marks a powerful moment in the life of a woman, because it definitely ends the time when living out the miracle of giving birth on the physical plane is possible. But if a woman can make peace with the biological end of this form of physical creativity, her creative energy can flow into the spiritual realm, to give birth to the divine child within, the Self.

My body would have waited longer to move into menopause without the effects of the chemotherapeutic drugs on my hormonal balance, but chemotherapy propelled me right over the threshold of this initiation doorway into the realm of the wise old woman.

On Vivien's birthday, the following insight came to me:

September 16, 1990, Dream:

Vivien is a baby, BEAUTIFUL. Someone asks me when I fell in love with her, and I say it was not really until this past summer.

Journal response to the dream:

So true, I did not know the depth of my own feeling for her until she decided to leave.

During this time, one year before discovery of the cancer, I continued to feel sick at not having my daughter there to nurture: I felt as if I were losing my inner child:

September 25, 1990, Dream:

I am transferring baby Vivien to the boat. She is on a tray and somehow slips and drops into the ocean on the other side of the boat. I see her and the tray sinking, both white.

I fix my eye on her and get ready to dive in, with open eyes, and go get her, to bring her back up quickly or it will forever be too late. Never mind if my contact lenses come out or I go down deep; she is vitally important to me! I wonder why she sinks and does not struggle to stay afloat.

The image of Vivien in the past two dreams was a beautiful blend of all the emotion attached to my love for my actual daughter and the divine child within, still incubating beneath my conscious level of awareness. The magic of this divine inner child was partly projected onto Vivien, and I felt as if I were losing my inner child when she left. My attachment to her *as my baby* needed to sink down into the ocean, back into the unconscious, both to free her to become an adult in my eyes and to free my energy for the coming spiritual work of birthing another aspect of the Self.

Images of pregnancy continued to come to me as I worked through the transition in my life from physical mother of another to physically mothering myself.

October 1, 1990, Dream:

I am seven months pregnant. I want to tell Dr. Krabill (the physician who performed the tubal ligation) about the failure of his operation.

Even though at a conscious level I did not yet know what this new life was about, the unconscious continued to present an image of pregnancy, which defied medical science.

Eventually, with the help of my analyst, I accepted Vivien's decision to leave and accepted Soli's marriage proposal. We got engaged in November of 1990, and following our engagement announcement, Vivien decided to come back and live with me again. She moved back in January of 1991, and I treasured the time with her.

Dreams of husbands, pregnancy, and babies continued as Soli and I planned our wedding and got married on April 14, 1991. Even getting the diagnosis of cancer on August 14, 1991, did not deter my dreams from continually giving me the message that new life was developing within me:

September 1, 1991, Dream:

I tell the dean of my doctoral program that I am pregnant. I feel my pants getting tight and feel the new life within me.

Then I joke that maybe instead of being pregnant, I skipped my period because I am going into menopause. I feel he will understand the complications of having a baby while doing intensive school work because he and his wife just had a baby too.

After talking with him, I cover up my beautiful breasts.

The dream shows that I had picked a school that was right for me, because it was a doctoral program in which the administration was sensitive to the new life within me because of the new life within them too. In reality, the school was extremely supportive of me as I juggled writing papers and flying out for classes with radiation, chemotherapy, and time out to care for myself. The end of the dream pictures me showing my nurturing side to the academic world, the union of two opposites within me, extroverted caring for others and introverted study.

September 25, 1991, Dream:
(during radiation therapy)

I am pregnant, only one more month till delivery, able to see my belly sticking out.

Soli is driving us in a Model-T Ford. In the car ahead of us are people from Pacifica, including my advisor, Dianne. The oracle said each woman would fulfill her work. For the women it was about growth, and for the men it was about "white." Not a good idea for the men to go into the Navy.

The dream indicated that the archetypal aspect of the pregnancy in me was about growth for women (and for the feminine principle within men), and the process was about "white" for the male principle. Since white light is a blend of all the colors in the spectrum, leaving nothing out, the dream might have been suggesting that the male principle needed to pull together all the different parts of the personality into a unified whole.

The dream advocated against going into an institution that is strictly regimented like the Navy, which is a color as well as the name of a branch of the armed forces. Any one color or one strict way of life would be too narrow for the Self, which wants all parts, all colors, working together to produce glorious white light.

At the point during the radiation treatments in Chapter 12 when I was physically ill and depressed from the side effects of the radiation, and Vivien and Soli were arguing over housekeeping issues, my grip on the inner vision of new life faltered a bit, and I wondered whether I would have a "miscarriage":

October 19, 1991, Dream:
(See Kirlian photograph 5.)

I am pregnant. I tell someone, "Here I am, forty-two years old and PREGNANT. I had a tubal ligation fifteen years ago and used another sterilization method, but here is a baby."

I say I am not sure the baby will make it full term. I ask about why it is there, and someone says there is a deep desire in me to give Soli a child. (It is his baby.)

In the dream, the conception was a double miracle: not only did the ovum bypass the tubal ligation, but it also outwitted a second sterilization method.

The Soli in the dream was my inner man, and the dream showed desire for an inner family, even though my weakened physical and emotional state presented the danger of not being able to carry the divine child to its full development.

As my energy picked up again during a break from the radiation treatments and I flew out to California for another session of classes, the baby motif continued:

October 25, 1991, Dream:

I walk a long way, holding my baby tender and close. I notice my left leg doesn't seem to pick up, because of the stress in the right one.

We pass Soli and hear him humming. We embrace. I'm so glad to see him. I ask, "Are we married yet?" He says yes, and I feel overjoyed. I feel my heart's desire is for him.

The dream showed connection with the inner male and joy at hearing that we were finally married. In the dream, the left side of the body, the right-brained unconscious side, was having trouble picking up things because of the stress in the right side, the conscious half of the body, which was going through severe physical stress from the cancer treatments. The dream brought up the delicate issue every person faces of how to join the opposites: left and right, academic life versus home life, male and female sides. Balancing demands from these opposing forces creates stress.

The course of this divine pregnancy was not always smooth, as shown by the trouble in the following dream:

November 29, 1991, Dream:

I take my six-month-old baby girl to the nursery to change her, and in the process, water accidentally gets on her head. She sputters, then says her first word, "water," enunciated clearly. I am delighted. Then she says other words, including "Hi!"

I take her outside to gather acorns to feed our salamander.

My baby girl is on a leash and falls down a step. To my utter dismay, I see the parts of her body have broken apart, and she can no longer function. The parts are dead.

Someone tells me, "You cannot put an initiation to the test," and I see where I went wrong.

Later in the dream she is ten months old, healthy and happy.

The point I see where I went wrong in the dream was putting the child on a leash; this action caused her to become dismembered. Animals are put on leashes to try to control their actions, but when my ego tried to keep the divine child on a leash in the dream, disaster followed.

The child of the Self is bigger than ego consciousness and cannot be controlled. The initiation needed freedom to develop along its own miraculous path (talking at only six months old); when tested or forced, it fell apart and died.

Yet this process of dismemberment was essential to the initiation. Old forms must be dismembered before they can be put back together into a new form. Without the death/dismemberment/crucifixion stage, no resurrection could follow. The Egyptian story of the body of the god Osiris being dismembered and the parts being strewn out all over the earth is a reflection of the feeling one has in the "deconstruction" stage of this process. Then Isis, the wife of Osiris, brought him back to life by gathering up all the parts and putting them back together again, the integration phase of the cycle. Ego control of my life needed to be dismembered so that regulation could be given over to the Self.

A further dream pointed out the dangers of trying to manipulate the birth:

December 16, 1991, Dream:

I get word the baby-sitter has died. Then I see the baby-sitter and her boyfriend. She is pregnant and trying to give birth. Her due date was in November.

She knows the baby inside has died. A nurse comes and pulls it out. As she does, she consults her cranial book on the type of delivery she has just made and looks up the appropriate skull pattern and visually memorizes the internal bone structures involved. I see them flashing like cross-section slides in her mind.

Then she does a cranial manipulation treatment to the newborn baby and it lives. I see pictures of the child at several years old and then again in adolescence. He sustains motor damage to the left side of his body, but he survived.

A baby-sitter is someone who cares for other people's children instead of her own, and as an inner figure, she needed to die in me so that I could focus all my energy on my own healing and on the birth of the divine child that was trying to gestate within me. However, the baby-sitter mentality popped back into life and propagated. The dream pointed out several errors that led to physical damage in the baby-sitter's offspring:

First, she was pregnant and *trying* to give birth, attempting to force a natural event that should be allowed to happen all by itself, in its own time.

Second, even though the baby was dead, the internal nurse manipulated it back into life, with the result that its left side was permanently damaged.

The process of birth was controlled too consciously, too much from the ego instead of from the Self; when life clearly leaves a baby, an inner potential we have been cultivating, the death needs to be respected as an act of God instead of trying to revive the dead baby. Likewise, if a mother forever clings to the baby stage of her child's life instead of allowing the baby aspect to die and transform into an adult, the child will be permanently damaged.

After the healing meditation on January 11, 1992, when I intuitively felt that the cancer had left my body, a beautiful baby image came in my dream world:

February 3, 1992, Dream:

At last my baby is born. I knew two months ago for certain that I was seven months pregnant, and now here she is.

She is beautiful! She sits with eyes wide open, and I communicate love to her through our eyes.

As I am riding with her on my bicycle, she nurses from my breast, eagerly and hungrily, and my milk flows well for her. However, I think about how I don't have enough hands for everything and perhaps she is not safe riding on the bike like this. I stop the bike with my feet.

At the baby care center the man in charge changes her diaper, and she has a bad diaper rash. I tell him I see I will need to change her more often.

What shall we call her? I am drawn toward the name "Chrisabelle," so she would be called "Chris" for short, like Christ.

Who is the father of this child? I try to think back who I had sex with nine months ago. Black? White? The only person I have had sex with in the last two years is Soli, so I think he must be the father.

Just after I felt the cancer in my body being released, this dream gave me the image of new birth, a beautiful child with wide-eyed, flowing love between the two of us. The child was clearly the divine child, named "Chrisabelle," like Jesus Christ, who was conceived by God and birthed by a human woman.

Christ's path along the road back home to God was not always smooth, just as initiations are difficult by their very nature since they require change and adaptation. Jung (1970) wrote about the process of transformation encountered by the initiate being similar to the passion of Christ:

> It is not an "imitation of Christ" but its exact opposite: an assimilation of the Christ-image to his own self. . . . It is no longer an effort, an intentional straining after imitation, but rather an involuntary experience of the reality represented by the sacred legend. (p. 349)

As Christ let go of his human form, allowing it to be sacrificed on the cross, he came into a deeper connection with the divine as he resurrected with an indestructible body, one that had died and was then transformed. Like the Christ story, when I allowed the roles my ego had been fixated on (giver but not receiver, never angry, person with hair on her head) to be sacrificed on the cross, then one by one they were transformed into new life.

In the dream, I realized that I had my hands full caring for Chrisabelle with everything else in my life, and I also realized that I would need to be conscious about her diapers needing changing, conscious of removing the parts of food leftover after the nourishment is extracted, parts which become toxic to the skin if left in the diaper. To make room in my life for her, I had to be conscious of eliminating parts of my life that no longer nourished me.

The ending of the dream reinforced the divine nature of the child by questioning her paternity; if the baby were a real pregnancy, I would have known who the father was. The dream raised the question of whether the father might be black or white. With light, white is a composite of all colors in the spectrum; however, with pigments, black is the result of a mixture of all the colors. Thus, black and white could be seen as two different forms of completeness, opposites that might have fathered the divine child. In the end, purely by logic, I deduce that Soli, my inner male, must be the father.

The following dream came three days before the fifth chemotherapy session, as I was gearing up for another round of bear-whalloping:

March 17, 1992, St. Patrick's Day Dream:

I am running in a long race. In a steep part going up a muddy hill into a forest area, a helper comes along, a black man or spirit man who gets immediately behind my body and helps me push myself up.

I get worn out and feel about to quit when I finish up the hard part. I continue running, and Vivien comes from the other direction. She shows me where the runners' lane is.

I ask her how far it is to the end, but she does not know. I feel irritated by her not knowing; however, she does know where the celebration is at the end. *I feel a coronation awaits me when I finish.*

I am holding a baby at the home of a woman who has a business and a family all together in one house.

Suddenly I notice the baby is no longer in my arms. Where is she? I look into the other room and see a young woman abusing the child by yanking on a rope around her neck, dragging her along the floor. The baby is screaming in pain and fright.

I quickly go and beat on the woman. I hit her in the head and kick her in the chest violently. Then I take the baby in my arms and calm her down. Others carry off the abusive woman, who is saying, "You don't know what they did to ME!"

I tell her, "You can't get the pain out of you by putting it on somebody else!"

The race mentioned in the dream could be referring to both the long, hard pregnancy and the marathon through chemotherapy. The path through both was hard and muddy, and I got tired. But a coronation awaited me at the end, a positive outcome of the whole process.

The second half of the dream could be seen on many levels. First, everyone has suffered at some place in life from the inner child not being allowed to be born and live its own life. Everyone has been yanked around somewhere, somehow.

On an obvious level, the dream presents a truth about the clinical picture of adults who actually abuse their physical children. These adults had been mistreated earlier, and since they could not tolerate the pain inside of themselves from this abuse, they tried to get the pain out of themselves by putting it into somebody else, a strategy that does not ultimately work.

On a deeper level, all the characters in the dream were also internal parts of me, and the woman who had the business and family all together may have been trying to crowd too many things together at one time in her life, with the result that the inner child got abused. I needed to be more realistic about what I could accomplish with only one of me, not crowding so many things into my life that I would not have time to care for my inner child.

I am both the woman yanking the child around by the neck, an avid picture of the businesswoman in me deciding to put my body through chemotherapy for "extra insurance" against cancer, and the frightened child screaming in pain, my physical body reacting against the needles and drugs.

This dream repeated the theme of the dream on November 29, 1991, by showing that pulling the child by a rope around her neck, a leash again, traumatized her emotionally and physically. Children need freedom to develop their unique personalities and are not supposed to be dragged around by over-controlling parents. Then another part of me beat against the abusive woman, rebelling against the demands my inner businesswoman put on me, and tried to comfort the inner child.

In the dream of the following night, a figure from the other side came to me, and I felt the importance of living in the present moment, a step in the right direction towards treating my inner child well:

March 18, 1992, Dream:

Vivien and Robin's grandmother, who died from cancer, leans over me and tells me she is now working as a "Spiritual Nurse."

She weeps as she tells me she sees that in the future a truck will spill its contents and poison the earth. I think to enjoy the present while we have it, and to also enjoy our journey to the other worlds when it comes.

I am playing with a little baby who is trying to learn to walk. Her legs are wobbly, but she is practicing trying to take a step. Cute!

Even though the earth may be headed for ecological disaster, the point of the dream was not to dwell on the negativity of that future event, but to live in the present, enjoying the beauty of the earth while we still have our planet. This attitude helped the inner child learn to walk, since the present is the only time we ever have.

Huge energy comes with the birth of the Self, but caring for the divine child also requires constant care. This baby is demanding, and keeping track of her was not always easy. I got sidetracked at times, as shown in the following dream:

April 4, 1992, Dream:

I am away from my baby girl for a couple days during class, but I return to her and she greets me warmly. Her name is something like "Noai." I talk with someone saying how I got pregnant again even after I had had a tubal ligation.

I put the baby at my breast and she nurses. I have a sugar-like substance on my nipples. I wonder if I will have an adequate milk supply after not nursing for two days.

I am traveling in my car and take a shortcut on foot through a store. Then I realize I need my car with me to continue the journey, and I backtrack.

Someone has taken my car with my baby daughter in it. I can get another car, but not another baby. I try to figure out how to find her again.

The theme of the baby's miraculous birth was repeated again, and this time her name was "Noai." Webster defines an "ai" as a sloth, so "no-ai" would mean that a slothful attitude is *not* right for caring for this divine child. She takes inner work and dedication.

"Noai" also sounds like "Noemi," the Spanish pronunciation for Naomi, the mother-in-law in the Biblical story[4] in the book of Ruth,

---

[4]Naomi had been living in the foreign country of Moab, and when her husband and two sons all died, she decided to return to her home country. Naomi told her two Moabite daughters-in-law to go back to the homes of their mothers; one stayed in her home country, but the other one, Ruth, chose to go with her mother-in-law.

who totally dedicated her life to Naomi, speaking these beautiful words:

> Entreat me not to leave you or to return from following you; for where you go I will go, and where you lodge I will lodge; your people shall be my people, and your God my God; where you die I will die, and there will I be buried. (Ruth 1: 16–17)

Caring for the divine child within, Noai, requires the same total and complete dedication, for life, that Ruth gave to Naomi.

As I settled into the period of lowered immunity between the last two chemotherapy rounds, the following dream came:

April 7, 1992, Dream:

I go into the delivery room to have my baby. The nurses and doctors are joking because I refer to the baby as "him." I say I really want a girl, but to avoid getting my hopes up falsely, I refer to the baby as masculine.

I feel dilation, and with the doctor's help I push the baby out, all without pain. She is a little baby girl.

I felt new birth within me as I finished the chemotherapy and began life anew, cherishing the gift of my life and taking time daily for meditation, yoga, and other forms of nourishing the physical body, the mind, and the soul. Although I realized that collective society as a whole would not support my need to nourish the Self with so much time and attention, "Chrisabelle" absolutely needed this care to grow. This new form of nourishing myself was pleasurable, not painful, just as this little baby girl was born miraculously, without pain.

The conflict between Vivien and Soli melted away as Vivien moved out into the world to an apartment of her own to study at a college two hours away from us. I had treasured the extra time with her as a bonus that came after I had already released her in my inner world, and when she left for the second time, I flowed with her decision instead of fighting it. I had Chrisabelle to care for and did not need Vivien to stay home in the nest past the time she should be out learning to fly.

Gradually I have been working through other areas of my life to improve the overall quality of my daily existence. I paced my doctoral studies at a rate I could handle and worked fewer clinical hours. I ate nourishing food, bought more purple clothes, treated myself to getting professional massage, took time to talk with friends and visit family, and continued with periodic acupuncture to keep my physical body in tune. I am in an ongoing process of making better friends with my physical body.

Though my dreams had been telling me for a long time to leave the stressful agency where I had been working, instead of quitting suddenly, I tapered off my caseload over the ten months following chemotherapy, while I regained my strength and laid the foundation for new professional contacts.

The day after I finally left that agency, I interviewed with Dr. Jon Kabat-Zinn (1990, 1994) of the University of Massachusetts Medical Center "Stress Reduction and Relaxation Program" and was invited to go through their internship program and participate in offering the program in Spanish in their Inner City satellite program.

The night before the first stress reduction internship class, I had the following dream, which showed a beautiful, healthy relationship with the divine child, who in this dream wore the face of my daughter:

February 9, 1993, Dream:

Vivien is there, a baby. She awakens and we grin at each other the first thing.

She is hungry and asks for milk. I feed her from my breast, and the milk comes out so fast she can't drink it all and spills some.

Then she takes her fill and then HUGS me. Oh, what a DELICIOUS EMBRACE, the two of us hugging chest to chest, the love flowing!

The baby in the dream was really Chrisabelle, symbolized by the close relationship I have with my actual daughter, Vivien, even though she is long past the precious time we had together when I breastfed her and is now out making her own life. The dream reflects the transformation the initiation of cancer brought to my life and shows the divine child within me being nurtured and LOVED.

Stephanie Simonton was also at this stress reduction internship class since 20/20 was filming us that day for a special on meditation (aired on February 19, 1993), and Stephanie said that the dream image she saw coming up most in her cancer patients was that of the crying, screaming baby, a dream figure I had often felt in my own dream world.

As I shared this dream with the internship group, I felt the joy of the embrace with this divine child flooding my soul and felt our love for each other filling my heart, a deeply satisfying sense of completeness.

I feel deep gratitude for the initiation of cancer, my teacher. Cancer began a process of intense healing of my body, my emotions, and my soul that is ongoing, my life's work. I lovingly accept the responsibility of daily caring for the divine child within me, attending to her physical, emotional, mental, and spiritual needs, because I love her. I accept the challenge of working towards balance in all parts of my life: the human little girl within and the adult, conscious and unconscious worlds, and masculine and feminine sides.

I love my body and my life. Cancer has brought me into a deeper relationship with the Self, the archetype of wholeness, and I am grateful for the experience.

I do not waste time asking, "Will I die?" Everybody is going to die sometime. Instead, I live in the moment and ask, "How can I live this very day, while I am still breathing, to the fullest?"

APPENDIX A

Cranial Osteopathic Philosophy

General Osteopathy

Osteopathy is an approach that takes the structure of both the physical and the non-physical body into account in its evaluation of a person's health. Osteopaths realize the importance of structure on the function of the body and use various modalities to affect structure, thus changing the body's function. Osteopathic doctors are also fully trained in all medical modalities, such as pharmacology and surgery, as practiced by allopathic medicine (which uses drugs to suppress symptoms of disease).

Cranial Osteopathy

A branch of osteopathic manipulation, cranial osteopathy, includes the structure of the cranial vault and thus the central nervous system in the evaluation and treatment process. The freedom of the sacrum (the end of the spine) to move in the pelvis is vital to the central nervous system, because the sacrum is attached to the cranial membranes through the dural sheath, the outer sheath of the spinal cord and brain.

All the fascia (the connective tissue) of the body is interconnected, so lesions anywhere in the body can affect a person's functioning. A cranial osteopath "reads" the fascia and other tissues to find out which parts of the body are troubled and then "tunes into" the body's own inherent self-corrective mechanism to help the body make whatever corrections that particular person needs. Approaching the inner physician is a spiritual endeavor, and corrections made by using the person's own healing mechanism can have far-reaching effects.

Cranial osteopathy is practiced by physicians because of their depth of knowledge of anatomy and physiology. No needles or drugs are used in this form of treatment—only concentrated attention and gentle pressure at times to the bones and tissues.

GWEN BROZ, D.O.
(Doctor of Osteopathy)

APPENDIX B

Kirlian Photography Theory

All living things—plants, animals and humans—not only have a physical body made of atoms and molecules, but also a counterpart body of energy. (Ostrander & Schroeder, 1971, p. 217)

This invisible counterpart, which interpenetrates the physical body, is slightly larger than the physical form. The energy has been called by many names throughout history, including the aura, the astral body, the subtle body, the etheric body, the emotional body, the fluidic body, the pre-physical body, the dreambody, and bio-plasma.

In the early 1900's, Dr. Walter Kilner discovered a method of actually seeing the aura around the human body by looking through glass goggles stained with coal tar dicyanin dye. He developed a system for diagnosing illness from his observations of the state of this energy field (Kilner, 1965).

This force field was observed by Soviet electrician Semyon Kirlian in 1939, when he noticed that the human body gives off a visible blue spark when coming in contact with a high frequency electrical field, and he set out to find a way to photograph this luminescence. He and his wife Valentina devoted their lives to researching this phenomenon, and their work was included in *Psychic Discoveries Behind the Iron Curtain* (Ostrander & Schroeder, 1971).

The Kirlians discovered that the life energy of a plant or human could be seen as visible colored light emanating from the organism, and the Soviet scientists called this energy field the "Biological Plasma Body," formed of "streams of masses of ionized particles" (Ostrander

& Schroeder, p. 216). Even when a portion of a living leaf had been cut away, the Kirlians found that *the energy body of the entire leaf showed up on the photo* (p. 214). American researchers trying to duplicate this "phantom leaf" effect have had more difficulty photographing this illusive phenomenon and estimate that with their Kirlian devices, only 5% of photographs show the phantom effect. However, three researchers working in India claim their repeatability rate for photographing the phantom leaf phenomenon is 50%. The experimenter must photograph the leaf within five minutes of the time the leaf is removed from the plant, or the phantom will disappear (Iovine, 1994).

Seventy percent of people who have had a limb amputated will continue to feel sensations from the missing arm or leg (Iovine). Although doctors explain this "phantom limb" phenomenon as nerves registering the amputated limb, clairvoyants have claimed to see the missing part in a "fluidic form," still attached to the body (Ostrander & Schroeder, p. 214).

In the 1930's, English medium Geraldine Cummins postulated that this invisible etheric body was "the link between the mind and the cells of the brain" (Ostrander & Schroeder, p. 215), and the Kirlian research supports the theory of the existence of this energy body, also thought to be the medium through which telepathic and clairvoyant phenomenon occur. The light that clairvoyants claim to see around the human body is the outer edge of this field.

This energy might also be related to the "morphogenetic field" which directs the growth of cells in the body, as in a salamander regenerating a leg that has been cut off, or our bodies healing cuts and scrapes. How else could each new cell formed at the site of a cut know whether to turn into skin, bone or other tissue unless some form of intelligence were directing the whole process?

To their surprise, the Kirlians also discovered that the energy pattern of a diseased leaf showed damage long before the damage was manifested in any physical change in the leaf, as if the energy body exists before the physical body and influences it, rather than the other way around.

On one occasion, Semyon Kirlian found that his equipment was giving chaotic energy pictures, and he assumed that his apparatus was malfunctioning. However, he soon discovered that the real reason he was not getting good photographs was that he was about to have an attack of a recurrent vascular illness he suffered from. The Kirlian device picked up the change in the field of his life force *before* he felt the onset of the attack.

After further research repeated the relationship of Kirlian energy pattern change to illness, leading Soviet medical professors reported, "Kirlian photography can be used for early diagnosis of disease, especially of cancer" (Ostrander & Schroeder, 1971, p. 211). Rumanian researchers issued a report in 1978 saying they had been able to detect breast cancer with 100% accuracy through Kirlian methods because cancer tumors photograph differently than normal tissue (Foreman & Hicks, 1989, p. 78). Rumanian medical researcher Dr. Ion Dumitrescu claimed that Kirlian photographs of unhealthy tissue show a bright image, whereas healthy tissue emits a dark image. He screened 6,000 chemical workers using a Kirlian technique and diagnosed malignant tumors in 47 of these workers. "Conventional tests confirmed 41 patients had malignant tumors. He feels conventional testing will confirm the other six workers as soon as their tumors become large enough to be detected by conventional means" (Iovine, p. 64).

Experiments done on plant cancers showed that plants with cancer repeatedly photographed with coronas that were more vivid than normal ones, because cancer raises the rate of metabolism in the plant (Moss, T., 1979), just like cancer raises the metabolic rate in humans (Quillin, 1994).[1] However, in tobacco plants with tumors, the sections of the leaves where the tumors were located *did not show up at all* on the Kirlian photographs (Moss, T., 1979). Cancerous tumors seem to be singularly devoid of Kirlian "life energy."

Kirlian photography has been used successfully to detect cystic fibrosis carriers, who produce a Kirlian corona that is uneven, broken, or clumped instead of a corona that is round and even (Iovine).

---

[1]Cancer installs abnormal biochemistry in its host, "changing the acid base balance, making the environment more favorable for cancer growth . . . blunting the immune system . . ." and "elevating metabolism and calorie needs while simultaneously lowering appetite and food intake to slowly starve the host" (Quillin, 1994, p. 1). Cancerous tumors love sugar and make the body with cancer crave sweets; the tumor then pulls most of the sugar from the blood into itself, not leaving enough for the rest of the body, so the craving for more sweets intensifies. The standard IV liquid diet fed to cancer patients is a high sugar, low protein formula, and this combination actually accelerates tumor growth. "A newly developed disease-specific formula of low sugar and high protein may selectively starve the tumor" (Quillin, p. 39). A sustained high-sugar diet like most Americans eat nourishes cancer cells and will eventually depress the immune system, since most of the immune bodies themselves are made of protein (Quillin).

Dr. Thelma Moss, a parapsychologist from UCLA, became interested in Kirlian photography when she read *Psychic Discoveries Behind the Iron Curtain* in the early 1970's. She traveled to Russia to visit the researchers in the book and was given some Soviet schematics for a few Kirlian devices. When she returned home, electronics experts said the device was unfeasible. However, an adult education student, Kendall Johnson, was able to rig a device from the schematics (Iovine). Moss and Johnson collaborated to study many topics, including acupuncture, faith healing, and bio-energy transfer. Both have published their findings (Moss, T. & Johnson, 1974; Johnson, 1975; Moss, T. 1979).

Researchers agree that the Kirlian photographs are not changed by the temperature of the fingertips or by dilation or constriction of the blood vessels. However, they disagree on whether the Kirlian photograph produced is related to Galvanic Skin Resistance.[2]

In separate studies, several psychiatrists have found that healthy people have a complete, mostly blue corona discharge from around their fingertips. In a schizophrenic patient, the corona is either unclear, patchy, or nonexistent. One study also found an unstructured red area under the schizophrenics' fingertips. Physical illnesses also produce corona changes with distinctive Kirlian patterns for gastrointestinal disturbance, upper respiratory infection, and psychiatric illness (Iovine).

Psychotherapist Lee Steiner (1977) used Kirlian photographs of right and left hand index fingertips to help diagnose the mental state of her clients. Steiner published many photographs showing that people with an emotional disturbance produce small, spotty Kirlian photo-graphs of their fingertips. On the other hand, persons who are emo- tionally healthy produce fingertip Kirlian photos of rich, full coronas. Steiner called this energy field "psychic energy" and postulated that "abundant psychic energy and emotional stability are

[2]Moss and Johnson claim that the Kirlian photographs do not portray sweat or Galvanic Skin Resistance (Moss, T. & Johnson, 1974). Iovine (1994) claims that when the Galvanic Skin Resistance drops, the overall corona discharge increases and that GSR drops under emotional stress. However, on the following page, he reports an Army study, which showed statistically valid conclusions that personnel who had been exercising (physical stress) produced larger-than-average corona patterns and that subjects under *mental stress* showed corona patterns that were smaller than average in diameter. Perhaps a drop in GSR does tend to increase the size of the corona, but maybe emotional stress also produces some other change in the energy field around the fingertip, which has a constricting effect that may override the increase from the drop in GSR.

identical" (p. 51). Steiner's book, titled *Psychic Self-Healing for Psychological Problems,* also recommends megavitamin supplements and a regular practice of entering a deep state of relaxation in meditation to pro-mote mental stability and to improve the corona pattern around the fingertips.

Russian researchers call this energy field the "bioplasmic body" and feel that it is replenished by the oxygen we breathe. "They relate this historically to the Indian Yoga philosophy, which explains that energy or 'Prana' is related to proper breathing. They also relate this to research that shows the positive effects in humans of negative-charged ionized air" (Iovine, p. 65).

The energy patterns of the five fingers move from the thumb being the most individual, representing ego, to the progressively more collective energy. The index finger shows domesticity and organization; the middle finger represents profession; the ring finger charts creativity, and the little finger stands for spirituality.

The left hand shows the pattern for the nonverbal, intuitive right brain, also thought to be the side of the unconscious mind in Jungian thought. The right hand stands for the left brain, the conscious, cognitive side of thinking. The left hand gets ideas, and the right hand puts them into reality.

Kirlian photography can be done without a camera. In a darkroom, the finger pads are placed on a piece of photographic paper on top of a metal plate. Then a high-frequency, high-voltage, low-amperage current is sent into this metal plate. The ions generated by the Kirlian device clash with the energy field of the human body, forming a spark in the visible blue and ultraviolet regions of the electromagnetic spectrum. The silver halide in the emulsion layer of photographic film is particularly sensitive to this light (Johnson, 1975). "Ion collisions yield enough energy to excite particles of air and make them luminous. This visible ionization has been termed corona, or brush, discharge and can sometimes be seen in darkness as a glow" (Johnson, p. 15).

When this paper is developed, it shows a negative image. Areas that had luminous sparks show up as black rays extending from the fingertips, with a white background. A contact positive can be made from this photo to show a black background with the spark areas showing as white areas of light.

APPENDIX C

Kirlian Apparatus

WARNING: Because of the high voltage used in this apparatus, electric shock can occur if not built and/or used properly. Anyone who does not have extensive technical knowledge of working with electricity should not attempt to build an apparatus from this design. The author and publishing company assume no liability for any damage done in any attempt to replicate this diagram. We are simply reporting the method that was used.

Under no circumstances should a live subject touch a ground while being photographed, or shock will result. For further safety, anyone who has a heart problem or a pacemaker should get the permission of their primary care physician before having a Kirlian photograph taken.

Iovine (1994) gives several diagrams of Kirlian devices that can be built, from simple up to more complex instruments.[1] He also gives detailed instructions on taking Kirlian photographs in color.

---

[1]Iovine also lists the following sources of complete Kirlian devices:

Mankind Research Ltd. Inc.
1215 Apple Ave.
Silver Spring, MD 20910
(301) 587-8686

Images Co.
P.O. Box 140742
Staten Island, NY 10314
(718) 698-8305

The apparatus that we used, shown in the following schematic, was housed in a plexiglass box with an on-off switch on the side. The back of the box had the capacitor switch, which could be set on either high, medium, or low. We needed to use the high setting to get enough power for photographs of the fingertips.

An additional piece of plexiglass was placed on top of the Kirlian box for insulation. Above the plexiglass was placed a 9 × 12 inch aluminum plate 1/4-inch thick that had been ground down to form a surface that was as smooth as possible. This aluminum plate was connected to the 6-volt battery by a conductor with alligator clips on both ends.

We used a red safelight so we could see what we were doing when we took these pictures. We placed unexposed black and white photographic paper, emulsion side up, on top of the aluminum plate. I then put my fingertips on the paper and turned on the switch for three seconds. (A longer exposure time will generate more sparks and darker pictures.) I also had to put a handle on the on-off switch so that I would not get a shock from turning it on. Another tip to avoid shock is to be sure to have no direct contact between the fingers and the aluminum plate. In addition, the fingertips must be at least a quarter of an inch away from the edge of the photographic paper, or a spark will jump right over the edge and give a mild shock.

Unlike the original schematic (Martin, 1974, p. 72), this unit operates on one 6-volt lantern battery B1. B2 is shown as a 22.5 VDC battery; however, they are not easy to purchase, so we supplied a power supply that converts 120V, 60 Hz to 20 to 24 VDC and used this as B2. Also, several other components were changed. These changes are noted on the schematic as well as the parts list.

PARTS LIST

Transistors:

Q1 & Q2, NTE121 or 2N554
Q3, uni-junction ECC5513 or 2N2646
D1 Diode IN4005

All *resistors* should be at least 1/2 watt:

2 39 ohm resistors
2 270 ohm resistors
1 5.6 K ohm resistor

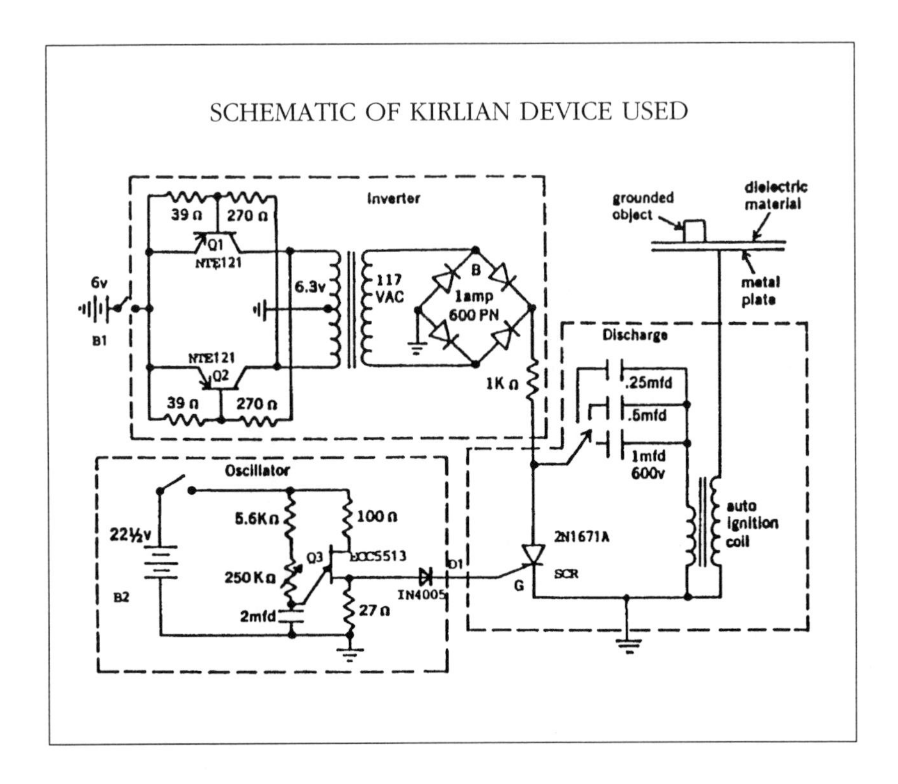

1   100 ohm resistor
1   27 ohm resistor
1   250 K ohm variable resistor

Bridge rectifier (B) should have 1 amp output.

6V auto *ignition coil*

6V lantern *battery*

22.5V *battery* or 1–120 volt, 60 Hz to 20–24 volt converter

Capacitors:
.25 MFD (600 V)
.5 MFD (600 V)
1.00 MFD (600 V)
2 MFD capacitor (50 V)

Toggle switch, 3 way, single pole

Plastic *Project Box* to mount in

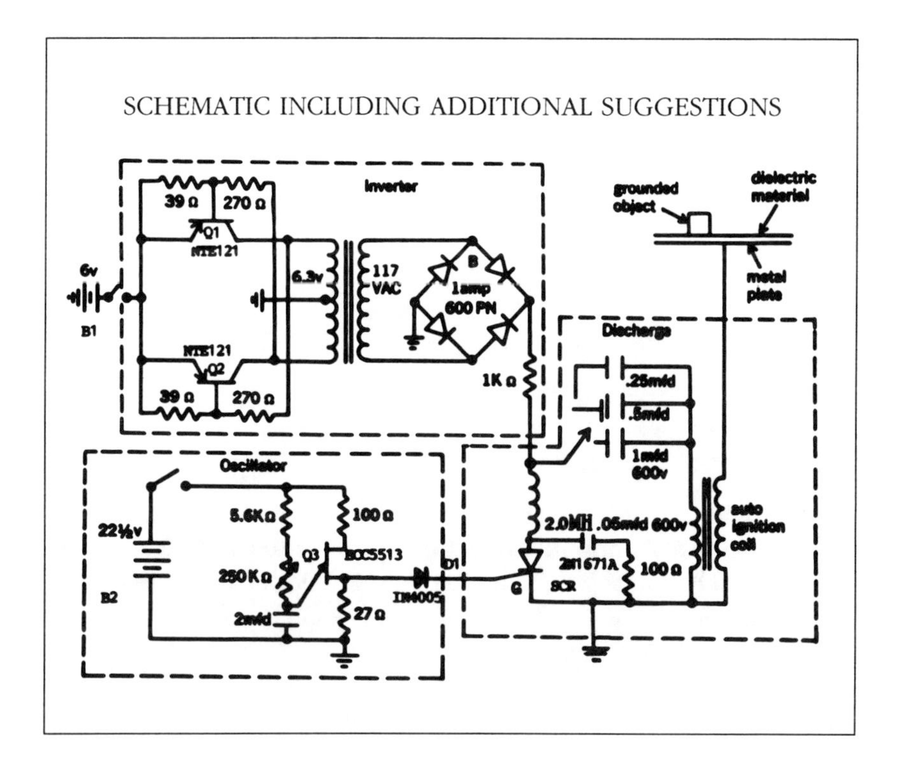

Metal plate 8" × 10" copper or aluminum, flat

2 regular size *alligator clips*

SPECIAL NOTES

As added protection in the discharge part of the schematic, diode D1 was added to protect Q3 from a reverse voltage spike. Also, to make the circuit work, the cathode on SCR 2N1617A had to be grounded.

To go into production on these units, the following components are recommended to be added:

1.) 2 millihenry coil

2.) .05 MFD capacitor, 600 V, in series with a 100 ohm resistor to protect the SCR.

3.) For safety, two on/off switches instead of one. This addition might cut the high voltage and cross-talk across the switch.

APPENDIX D

Acupuncture

Introduction

You may already know that the Chinese invented gunpowder and were advanced astronomers, but did you know that they discovered circulation and that the heart was the body's circulatory pump? They also inoculated smallpox victims by having them inhale the powder from the pustules of victims far before William Jenner did his famous cowpox inoculation experiments with milkmaids. Asian knowledge of the pharmacological properties of herbs and herbal combinations was extremely sophisticated. Asian medicine formulated an integrated, comprehensive system of pharmacopoeia by the fourth century A.D., a system which continues to be used and refined today.

The Theory of Asian Medicine

The main goal of acupuncture and Asian medicine is to bring the body into balance and harmony. Acupuncture works with energy to stimulate the body's inherent healing mechanism, which is called "chi."

Although at first glance, the concept of *chi* may sound metaphysical, we see evidence of *chi* everywhere. In the realm of physiology, it corresponds with biomagnetic energy. For instance, we understand from a Western perspective that the body is permeated with biomagnetic fields that are in a constant state of flux. We utilize EKG's and EEG's as diagnostic indicators of biomagnetic activity, and ohmmeters have been used to corroborate the traditional locations of acupuncture points on the surface of the skin.

From a broader perspective, *chi* is all around us. It is what stimulates growth; it is in the air that we breathe; it is the difference between life and death. Westerners can start from this framework to consider *chi* and how it operates in the healing mechanism of acupuncture.

The nervous system and the circulatory system are marked by networks and pathways through which flow impulses and blood; likewise, *chi* circulates throughout the body's networks through what we call meridians. These meridians, or energy channels, flow on the body's surface in a continual rhythmic cycle.

Stimulation of certain acupuncture points on the meridians can affect a problem either on the surface of the body, in the internal organs, or both. Acupuncture points are therefore channels of communication to the function of the internal organs. Each point produces different actions and effects on the body. Studies have shown that the stimulation of certain acupuncture points can produce changes in physiological functions such as increasing endorphin and serotonin levels, increasing cortisol levels, changing triglyceride and cholesterol levels, changing other endocrine levels, and changing digestive enzymes and antibody levels.

Diagnosis

So what is the acupuncturist's approach to the patient who comes into the office? Diagnosis entails a full case history, which includes questions about every aspect of a patient's functioning: the taste a patient may often crave, sleep requirements, elimination patterns, menstrual flow, appetite, and quality of pain experienced. Other important factors also include the patient's most favored or detested season and moods or emotional states that may often predominate in a patient's experience. In addition to the information gathered from these questions, the acupuncturist also feels the pulses at several locations on the wrist and examines the tongue. Tongue diagnosis and pulse diagnosis are an art in Asian medicine. Abdominal palpitation may also be part of the diagnosis.

All of this information should point out the disharmony present in the system. Patterns of disharmony are clinical syndromes as the Asians observed them. By understanding the pattern in a collection of symptoms, we can address the underlying imbalance that is contributing to the individual's health problems. This information determines treatment protocol. The correct treatment may include any of the following: proper acupuncture point selection, choice of needles, magnets, sometimes cold laser, warming of the needles with an herb called moxa, or stimulation of the needle with a low voltage electrical current.

The Complimentarity of Asian and Western Medicine

Acupuncturists coordinate their services with other health care professionals. Asian and Western medicine are very different systems, and each has its particular strengths. The optimum care for the individual comes from the integration of the two systems, a practice that is increasing in many parts of the world. In China, practitioners of Eastern and Western medicine work side by side in the hospitals. The patient can choose a track that would involve just Western, just Eastern, or a combination of the two systems.

Most people are familiar with the common application of acupuncture in Chinese hospitals as the main form of anesthesia employed and as a way to relieve post-operative pain. Acupuncture is also commonly used in the United States for pain relief. However, Asian medicine also addresses a multitude of additional health concerns. Since Asian medicine is an intervention that promotes homeostasis, it is particularly appropriate and effective with many of the frustrating disorders that involve dysfunction of the immune system, such as Crohn's disease, rheumatoid arthritis, chronic fatigue syndrome, AIDS, multiple sclerosis, lupus, and chronic colds and flus. Acupuncture is also effective in treating the side effects of oncology treatments and helps to boost the function of the immune system.

KATHLEEN TAMILIO-AWED,
Licensed Acupuncturist
President of Massachusetts Association
of Acupuncture and Oriental Medicine

APPENDIX E

Funding for Breast Cancer Research

This appendix compares the budget allotments for prostate cancer, which is found exclusively in males, to the budget allotments for breast cancer, whose incidence is 99% female.

An article in *Newsweek* stated that public awareness of prostate cancer research was "twenty years behind" breast cancer and cited statistics that funds for breast cancer research had more than tripled in the previous three years, with the 1994 National Institute of Health grants for breast cancer coming to $299 million while the allotment for prostate cancer research totaled only $51 million (Adler, Rosenberg, and Springen, 1993, p. 40).

However, a more critical look at the situation might view the facts differently. The National Cancer Institute (NCI) is one leg of the government's National Institute of Health (NIH), and NCI funding for breast cancer research has indeed more than tripled in the past four years. But their funding for prostate cancer research has more than quadrupled:[1]

NCI Funding for Cancer Research (in millions)

Year	Breast	Prostate
1990	$ 81.0	$13.2
1992	$145.0	$31.4
1994	$263.0	$55.0

[1]These figures came from a phone call to the NCI Budget Office, (301) 496-5803, February 10, 1994.

Also, the NIH and the NCI are not the only funding sources for cancer research. A review of the *Directory of Biomedical and Health Care Grants 1992* showed that the estimates allotted for prostate cancer research in fiscal year 1991 were $285,107,000 (1991, p. 188), and the allotments estimated for breast cancer research were $211,240,000 with an additional smaller grant of $22,905,000 for studying mammograms and risk factors for breast cancer (p. 80). These figures bring the total for breast cancer to about $234 million, versus $285 million for prostate cancer.

Another factor to consider in comparing these two forms of cancer is that breast cancer affects women differently than prostate cancer affects men. While prostate cancer killed around 35,000 American men in 1993 and breast cancer killed 46,000 women, prostate cancer tends to occur later in life than breast cancer. The median age for death from prostate cancer is age 77 (Adler, Rosenberg, and Springen, 1993, p. 40), which is five years older than the average life span of the American male. Doctors frequently choose not to treat prostate cancer because old age may kill the victim first: doing nothing is often the best treatment for prostate cancer.

Breast cancer hits an average of roughly fifteen years earlier than prostate cancer; since breast cancer generally strikes earlier, it has more time to destroy quality and quantity of life. "Women who die of breast cancer lose an average of nineteen years of life" according to the Massachusetts Breast Cancer Coalition's Breast Cancer Fact Sheet (1992). Doing nothing is *not* usually the treatment of choice for breast cancer.

Prostate cancer has shown a slow, steady rise in the death rate of males from 1930 to 1985. The death rate for lung cancer has shown an extremely sharp rise in males since 1930 and passed the death rate for prostate cancer in 1949. The form of cancer that is the number one killer of white males is lung cancer. The form of cancer that up until now has been the number one killer of white females has been breast cancer (National Cancer Program, 1983–1984, pp. 5–6). In the United States 58,000 men and women died in the ten-year Vietnam War; 330,000 women died of breast cancer over this same time period (Breast Cancer Fact Sheet, 1992).

The death rate from breast cancer has held quite steady since 1930. The death rate of lung cancer in women has been rising over this same time period. It had not yet caught up with the rates for breast cancer by 1981 (National Cancer Program, p. 6). However, the

1994 statistics put out by the American Cancer Society showed that lung cancer has now surpassed breast cancer as the leading cancer killer of women, with 53,000 deaths per year from lung cancer compared to the 46,000 deaths per year from breast cancer. The American Cancer Society also noted that the deaths from lung cancer could be reduced by 40,000 per year if women just did not smoke.

Although Black and Hispanic females have a slightly lower incidence of breast cancer than white females, breast cancer was still the type of cancer that had the highest incidence in all three groups of women in the statistics from 1984. Black males have a significantly higher rate of both lung cancer and prostate cancer than the rate for these types of cancer in white males (National Cancer Program, p. 9).

A breakdown of funding by nonprofit organizations and foundations for cancer-related activities in 1984 showed a total funding of $222,872,000 with $10,800,000 donated to the American Lung Association and $2,394,000 given to the Council for Tobacco Research, both areas connected to lung cancer, which has long been the number one cancer killer of men. A total of $3,000 was donated to the Breast Cancer Advisory Center, the only money marked for breast cancer, which at the time was the number one cancer killer of women. In comparison, $14,000,000 was given to The Leukemia Society of America and $485,000 went to the Skin Cancer Foundation (National Cancer Program, p. 45).

In the late 1980's a company called Hybritech marketed a blood test for prostate cancer. Since a cancerous prostate tends to release a protein produced exclusively by prostate cells into the bloodstream, a rise in the level of PSA (prostate-specific antigen) in the blood is "a good rough indicator of tumor activity" (Cowley, Springen, Rosenberg, and Ramo, p. 42). During my treatment, no blood marker for breast cancer was available. However, since that time a blood test has been developed that is a general indicator for monitoring the level of tumor activity in women who have already been diagnosed with breast cancer.

Relatively little information has been published on recovery from breast cancer. A review of Medline (a computer database for medicine) from 1985 to June of 1992 found only twenty-seven articles in the medical literature on recovery from breast cancer. Twenty-five of these articles dealt strictly with the efficacy of specific chemotherapy agents. One talked about suppression of the immune system in irradiated breast cancer patients, and the remaining article was written on the compliance of breast cancer patients with their chemotherapy regimen.

The only article in the medical literature from this time period on *prevention* of breast cancer (Goodman & Goodman, 1986) took a quite critical viewpoint of all forms of prevention efforts in place at the time but did not offer alternative prevention methods. For example, the article cited statistics that ran counter to the generally accepted principle that eating a low-fat diet aids in preventing breast cancer, pointing out some European countries that had a higher fat diet but lower rates of breast cancer than their neighbors. The article also criticized medical doctors for viewing total hysterectomy as a viable method to prevent breast, ovarian, and uterine cancer; the authors reasoned that although male castration reduces the rates of genital-related cancers in men, doctors do not routinely prescribe prophylactic male castration. The article also criticized some health leaders for launching mass prevention programs with no sound biomedical foundation, saying sometimes magical powers were attributed to certain symbols. The authors wanted scientific studies to back up claims that alternative prevention methods actually worked. A further criticism was leveled at those who transferred responsibility for the disease to the victim without mentioning that for the person who already has a diagnosis of cancer, understanding that the disease might be influenced by an emotional pattern *that can be changed* gives the patient the possibility of having some control over the outcome of the illness and prevents a feeling of being totally helpless in the face of a powerful disease.

References

Adler, J., Rosenberg, D., & Springen, K. (1993, December 27). The Killer We Don't Discuss. *Newsweek,* pp. 40–41.

Amaa-Ra, S. A. (1991, Summer). *The Starry Messenger,* p. 1.

Anthony [El Saffar], R. (1994). *Rapture Encaged—The Suppression of the Feminine in Western Culture.* New York: Routledge.

Associated Press. (1992, October 19). Rally targets breast cancer. *Telegram & Gazette,* p. 1 and 4.

Becker, R. O., & Selden, G. (1985). *The Body Electric: Electromagnetism and the Foundation of Life.* New York: William Morrow and Co., Inc.

Benson, H. (1975). *The Relaxation Response.* New York: Avon Books.

Blasdell, K. (1989). Acute Immunoreactivity Modified by Psychosocial Factors: Type A/B Behavior, Transcendental Meditation, and Lymphocyte Transformation (Doctoral dissertation, Maharishi International University, 1989). *Dissertation Abstracts International, 50,* Issue 10, Section B.

Borysenko, J. (1988). *Minding the Body, Mending the Mind.* New York: Bantam Books.

Campbell, J. (1968). *The Hero with a Thousand Faces* (2nd ed.). Princeton, NJ: Princeton University Press.

Chopra, D. (1991). *Perfect Health: The Complete Mind/Body Guide.* New York: Harmony Books.

———. (1988). *Return of the Rishi: A Doctor's Search for the Ultimate Healer.* Boston: Houghton Mifflin Co. (The 1991 edition is titled *Return of the Rishi: A Doctor's Story of Spiritual Transformation and Ayurvedic Healing.*)

Cohen, D. (1991). *The Circle of Life: Rituals from the Human Family Album*. Hong Kong: Harper San Francisco.

Colbin, A. (1986). Food and Healing. New York: Ballantine Books.

Cooper, J. (1978). *An Illustrated Encyclopaedia of Traditional Symbols.* London: Thames & Hudson Ltd.

A Course in Miracles. (1976). Glen Ellen, CA: Foundation for Inner Peace.

Cousins, N. (1979). *Anatomy of an Illness as Perceived by the Patient: Reflections on Healing and Regeneration.* New York: W. W. Norton & Co.

———. (1991). *Head First: The Biology of Hope and the Healing Power of the Human Spirit.* New York: Penguin Books.

Cowley, G. (1993, December 6). The Hunt for a Breast Cancer Gene. *Newsweek,* pp. 46–49.

Cowley, G., Springen, K., Rosenberg, D., & Ramo, J. C. (1993, December 27). To Test or Not to Test. *Newsweek,* pp. 42–43.

Directory of Biomedical and Health Care Grants 1992 (6th ed.). (1991). Phoenix: Oryx Press.

Foreman, L., & Hicks, J., eds. (1989). *Powers of Healing.* Alexandria, VA: Time-Life Books.

Foster, S., & Little, M. (1987). The Vision Quest: Passing from Childhood to Adulthood. In L. Mahdi, S. Foster, & M. Little, eds., *Betwixt & Between: Patterns of Masculine and Feminine Initiation* (pp. 79–110). La Salle, IL: Open Court.

von Franz, M. L. (1964). The Process of Individuation. In C. G. Jung, ed., *Man and His Symbols* (pp. 158–229). New York: Anchor Press.

———. (1972). *The Feminine in Fairy Tales.* Zürich: Spring Publications.

———. (1975). *C. G. Jung: His Myth in Our Time.* Boston: Little, Brown & Co.

Freud, S. (1959). Mourning and Melancholia. In J. Riviere, trans., *Collected Papers* (Vol. 4, pp. 152–170). New York: Basic Books, Inc. (Original work published 1917)

Fried, M. H., & Fried, M. N. (1977). *Transitions: Four Rituals in Eight Cultures.* New York: Norton & Co.

Fritchie, R. G. (1987). *A Path to Self-Healing Using God's Love and Crystals.* Unpublished manuscript.

van Gennep, A. (1960). *The Rites of Passage* (M. B. Vizedom & G. L. Caffee, trans.). Chicago, The University of Chicago Press. (Original work published 1908)

Gibran, K. (1966). *The Prophet*. New York: Alfred A. Knopf.

Goleman, D., & Gurin, J., eds. (1993). *Mind/Body Medicine: How to Use Your Mind for Better Health*. New York: Consumer Report Books.

Goodman, L. E., & Goodman, M. J. (1986, April). Prevention—How Misuse of a Concept Undercuts its Worth. *Hastings Center Report, 16* (2), 26–38.

Gorbach, S. L., Zimmerman, D. R. , & Woods, M. (1984). *The Doctors' Anti-Breast Cancer Diet*. New York: Simon & Schuster.

Grinspoon, L., & Bakalar, J. B. (1993). *Marihuana, the Forbidden Medicine*. New Haven, CT: Yale University Press.

Halifax, J. (1982). *Shaman: The Wounded Healer*. New York: Thames & Hudson.

Hamilton, E. (1942). *Mythology*. Boston: Little, Brown & Co.

Hay, L. (1985). *You Can Heal Your Life*. Santa Monica, CA: Hay House.

Henderson, J. (1964). Ancient Myths and Modern Man. In C. G. Jung, ed., *Man and His Symbols* (pp. 104–157). New York: Anchor Press.

Iovine, J. (1994). *Kirlian Photography: A Hands-On Guide*. Blue Ridge Summit, PA: TAB Books.

Johnson, K. (1975). *Photographing the Nonmaterial World*. New York: Hawthorne Books.

Jung, C. G. (1968). *The Archetypes and the Collective Unconscious* (2nd. ed.). Princeton, NJ: Princeton University Press.

———. (1970). *Mysterium Coniunctionis*. Princeton, NJ: Princeton University Press.

———. (1974). *Dreams*. Princeton, NJ: Princeton University Press.

Jung, C. G., ed. (1964). *Man and His Symbols*. New York: Anchor Press.

Kabat-Zinn, J. (1990). *Full Catastrophe Living: Using the Wisdom of Your Body and Mind to Face Stress, Pain, and Illness*. New York: Delta.

———. (1994). *Wherever You Go, There You Are*. New York: Hyperion. Princeton, NJ: Princeton University Press.

Karpinski, G. (1990). *Where Two Worlds Touch*. New York: Ballantine Fawcett Trade Books.

Kemeny, M. (1993). Emotions and the Immune System. In B. Moyers, *Healing and the Mind* (pp. 195–211). New York: Doubleday.

Kilner, W. (1965). *The Human Aura*. New York: University Books.

Krippner, S., & Rubin, D. (1974). Kirlian Photography, Acupuncture, and the Human Aura: Summary Remarks. In S. Krippner & D. Rubin, eds., *The Kirlian Aura: Photographing the Galaxies of Life* (pp. 181–187). Garden City, NY: Anchor Press/Doubleday & Co., Inc.

Lappé, F. M. (1975). *Diet for a Small Planet* (rev. ed.). New York: Ballantine Books.

Lauder, E., ed. (1991, October). Self's Breast Cancer Report. *Self,* pp. 135–145.

LeShan, L. (1977). *You Can Fight for Your Life: Emotional Factors in the Treatment of Cancer.* New York: M. Evans and Co., Inc.

Lockhart, R. (1983). *Words as Eggs.* Dallas, TX: Spring Publications, Inc.

Love, S. (1990). *Dr. Susan Love's Breast Book.* Reading, MA: Addison-Wesley Publishing Co.

———. (1993, May/June). Confronting Breast Cancer: An Interview with Susan Love. *Technology Review,* pp. 45–53.

McGarey, W. (1988). *Healing Miracles: Using Your Body Energies.* San Francisco: Harper & Row.

Markman, M., Theriault, R., & Williams, P. A. (1991). Making Cancer Chemotherapy More Tolerable. *Patient Care, 25,* 37–55.

Martin, R. (1974). A Portable Kirlian Device. In S. Krippner & D. Rubin, eds., *The Kirlian Aura: Photographing the Galaxies of Life* (pp. 72–74). Garden City, NY: Anchor Press/Doubleday & Co., Inc.

Moss, R. (1991). *The Cancer Industry.* New York: Paragon House.

Moss, T. (1979). *The Body Electric: A Personal Journey into the Mysteries of Parapsychological Research, Bioenergy, and Kirlian Photography.* Los Angeles: J. P. Tarcher, Inc.

Moss, T., & Johnson, K. (1974). Bioplasma or Corona Discharge? In S. Krippner & D. Rubin, eds., *The Kirlian Aura: Photographing the Galaxies of Life* (pp. 51–71). Garden City, NY: Anchor Press/Doubleday & Co., Inc.

Moyers, B. (1993). *Healing and the Mind.* New York: Doubleday.

National Cancer Program. (1983–1984). *Director's Report and Annual Plan FY 1986–1990.* National Institutes of Health, U.S. Department of Health and Human Services.

Ostrander, S., & Schroeder, L. (1971). *Psychic Discoveries Behind the Iron Curtain.* Englewood Cliffs, NJ: Prentice-Hall.

Quillin, P. (1994). *Beating Cancer with Nutrition.* Tulsa: The Nutrition Times Press, Inc.

Ragland, D. R., & Brand, R. J. (1988). Type A Behavior and Mortality from Coronary Heart Disease. *New England Journal of Medicine, 318,* 65–69.

Robbins, J. (1987). *Diet for a New America.* Walpole, NH: Stillpoint Publishing.

Selye, H. (1976). *The Stress of Life.* New York: McGraw-Hill.

Siegel, B. (1986). *Love, Medicine, & Miracles.* New York: Harper & Row, Inc.

Silbey, U. (1986). *The Complete Crystal Guidebook.* San Francisco: U-read Publications.

Simonton, C., Matthews-Simonton, S., & Creighton, J. (1978) . *Getting Well Again.* New York: Bantam Books.

Simonton, C., & Simonton, S. (1975). *Mind As Healer, Mind As Slayer: Belief Systems and Cancer* (Cassette Recording).

Simonton, S. (1993, June). Can Psychotherapy and Life-Style Changes Affect the Outcome of Cancer? *Cancer and Healing.* Symposium at the Interface Summer Conference, Cambridge, MA.

Snellgrove, B., & Snellgrove, M. (1979). *The Unseen Self: Your Hidden Potential.* London: Kirlian Aura Diagnosis.

Spielberg, S. (Producer and Director). (1985). *The Color Purple* (Film). Burbank, CA: Warner Bros.

Spletter, M. (1982). *A Woman's Choice: New Options in the Treatment of Breast Cancer.* Boston: Beacon Press.

Steiner, L. R. (1977). *Psychic Self-Healing for Psychological Problems.* Englewood Cliffs, NJ: Prentice-Hall.

van Tets, W. F. , Leenen, L. P., Roukema, J. A., & Pipers, F. M. (1990). Bilateral Primary Breast Carcinoma in a Man. *Netherlands Journal of Surgery,* 42(6), 158–160.

Turner, V. (1987). Betwixt and Between: The Liminal Period in Rites of Passage. In L. Mahdi, S. Foster, & M. Little, eds., *Betwixt & Between: Patterns of Masculine and Feminine Initiation* (pp. 3–19). La Salle, IL: Open Court.

Upledger, J. E. & Vredevoogd, J. D. (1983). *Craniosacral Therapy.* Seattle, WA: Eastland Press.

Walker, B. G. (1983). *The Woman's Encyclopedia of Myths and Secrets.* San Francisco: Harper & Row.

Webster. (1965). *Webster's Seventh New Collegiate Dictionary.* Springfield, MA: G. & C. Merriam Co.

Weiss, B. L. (1988). *Many Lives, Many Masters.* New York: Simon & Schuster.

Wilber, K. (1991). *Grace and Grit: Spirituality and Healing in the Life and Death of Treya Killiam Wilber.* Boston: Shambhala.

Wiley, W. (1992, October 19). Survivors Remember Women Who Fought, Lost. *Telegram & Gazette,* p. 4.

Woolger, R. J. (1987). *Other Lives, Other Selves: A Jungian Psychotherapist Discovers Past Lives.* New York: Doubleday.

The World Book Encyclopedia. (1977). Chicago: Field Enterprises, Inc.

Yatri. (1988). *Unknown Man.* New York: Simon & Schuster Inc.

Zoja, L. (1989). *Drugs, Addiction and Initiation: The Modern Search for Ritual.* Boston: Sigo Press.

Index

Audio Tape "Initiation" © and Workshop Information

The author can be contacted at the following address for lecture and workshop information and to order tapes:

Barbara Stone, Ph.D.
P.O. Box 12
Worcester, MA 01614

"Initiation"© is an audio tape which can be used for relaxation and meditation. The author's voice combines with solo piano music, composed and performed by the author, to assist in reaching a deep state of relaxation.

Side 1: Guided 36 minute journey through the body, relaxing part by part.

Side 2: Guided 20 minute sitting meditation, which can be extended for the entire 45 minutes.

Tapes are $10 @ plus $1 @ for postage and handling = $11 US each.

Massachusetts residents add 5% (50 cents) sales tax per tape.

To order tapes, please send a check or money order made out to Barbara Stone, Ph.D., to the above address.